The Great Cyclone at
St. Louis and East St. Louis,
May 27, 1896

Shawnee Classics
A Series of Classic Regional Reprints for the Midwest

Personal Memoirs of John H. Brinton
Civil War Surgeon, 1861–1865
John H. Brinton

Stagecoach and Tavern Tales of the Old Northwest
Harry Ellsworth Cole
Edited by Louise Phelps Kellogg

"Black Jack"
John A. Logan and Southern Illinois in the Civil War Era
James Pickett Jones

A History of the Ninth Regiment Illinois Volunteer Infantry,
with the Regimental Roster
Marion Morrison

The Outlaws of Cave-in-Rock
Otto A. Rothert

A Woman's Story of Pioneer Illinois
Christiana Holmes Tillson
Edited by Milo Milton Quaife

Army Life of an Illinois Soldier
Including a Day-by-Day Record of Sherman's March to the Sea
Charles W. Wills

THE GREAT CYCLONE

AT

ST. LOUIS AND EAST ST. LOUIS,

MAY 27, 1896.

BEING A FULL HISTORY OF THE MOST TERRIFYING AND DESTRUCTIVE
TORNADO IN THE HISTORY OF THE WORLD,

—WITH—

NUMEROUS THRILLING AND PATHETIC INCIDENTS AND PERSONAL
EXPERIENCES OF THOSE WHO WERE IN THE
TRACK OF THE STORM.

—ALSO—

AN ACCOUNT OF THE WONDERFUL MANIFESTATIONS OF SYMPATHY FOR THE AFFLICTED IN
ALL PARTS OF THE WORLD.

COMPILED AND EDITED BY

JULIAN CURZON,

THE DISTINGUISHED AUTHOR AND LITERATEUR.

SPLENDIDLY ILLUSTRATED

WITH 150 ACCURATE PHOTOGRAPHIC VIEWS AND ENGRAVINGS, PORTRAYING IN VIVID REALISM
THE WRECK AND RUIN WROUGHT BY THE DEMON OF DESTRUCTION.

Southern Illinois University Press
Carbondale and Edwardsville

Foreword by Tim O'Neil, copyright © 1997 by the Board of Trustees, Southern Illinois University. First published 1896 by Cyclone Publishing Company. All rights reserved. Printed in the United States of America.

00 99 98 97   4 3 2 1

Library of Congress Cataloging-in Publication Data

The great cyclone at St. Louis and East St. Louis, May 27, 1896 : being a full history of the most terrifying and destructive tornado in the history of the world . . . / compiled and edited by Julian Curzon.

p. cm.

Originally published: St. Louis : Cyclone Pub., 1896.

1. Tornadoes—Missouri—Saint Louis—
History—19th century.
2. Saint Louis (Mo.)—
History. I. Curzon, Julian.
F474.S257G74 1997
977.8 ' 6503—dc21
ISBN 0-8093-2124-6
(alk. paper)
96-29716
CIP

The paper used in this publication meets the minimum requirements of American National Standard for Information Sciences—Permanence of Paper for Printed Library Materials, ANSI Z39.48-1984. ∞

CONTENTS.

	PAGE.
List of Illustrations	xi
Foreword by Tim O'Neil	xv

PUBLISHERS' NOTICE	15 - 16
INTRODUCTORY	17 - 20
The Great Cyclone	21
A Discriminating Storm	22
The Cyclone's Fatal Path	25
All the Elements of a Tornado	27
Warnings of Deadly Peril	31
Gathering Force in its Course	32
On the Outskirts of the Storm Region	34
The Full Force Felt	38
Where Deaths were Most Frequent	40
The Attack on the River	43
The River a Raging Torrent	45
Eleven Ladies in Deadly Peril	49
The Damage in East St. Louis	51
A Jury's Narrow Escape	52
Predicted a Year in Advance	56
Barometrical Readings	57
The Text of the Forecast	61
How Loss Could Have Been Lessened	63
Death in the Wind	67
Cyclones of History	68
A Landmark in Irish Records	70
Storms in Our Own Land	73
Victims of the Cyclone	75
Surveying the Wreck	76
A Suburb Desolated	80
In an Older Section	85
Mourning the Ruin of the Park	87
Homes and Churches Destroyed	91
In Old St. Louis Proper	93
The City Hospital Wrecked	94

CONTENTS.

	PAGE.
Removing the Sufferers	97
Physicians' Heroism	99
Horrors at the Poor House	105
Providential Escapes	109
Union Depot Power House	111
A Marvel of Re-Construction	115
Street Railway Service Paralized	117
Hours of Horrible Suspense	118
Telegraph Wires Down	122
A Fated Corner	123
A Combination of Horrors	128
Handsome Residences Destroyed	130
Saw Her Children Burn	134
Homeless and Destitute	135
Horses Killed in Harness	139
Desolation and Destruction	140
Losses to Business Concerns	141
Swamped By Their Neighbor	142
Tremendous Damage	145
Damage to Homes on One Street	146
Around the City Hospital	148
Many Dead in a Wrecked Territory	151
Nothing was Spared	152
Searching for Their Husbands	154
Gas Works Wrecked	157
Fallen Buildings Everywhere	158
Union Club Wrecked	159
Lives Lost in a Fire	160
A Huge Factory Destroyed	163
Churches Wrecked	166
A Quarter of a Million	169
Struck by the Tornado	170
An Appalling Loss	171
Damage East of Broadway	172
Panic Among Factory Girls	176
Devastation's Awful Triumph	177

CONTENTS.

	PAGE.
Station Torn up	181
In North St. Louis	182
Panic at the Races	183
Rescued from a Wrecked Saloon Building	188
Down-Town Wrecks	189
Chamber of Commerce Unroofed	193
Skyscrapers Escaped	195
At the Court House	196
At the Custom House	199
Scenes at the Hotels	199
Danger at the Jail	201
Heartbreaking Spectacle	205
Bringing in the Dead	207
In East St. Louis	211
Prayed for Mercy	212
Searching for the Dead	214
A Hotel's Sad Role	214
Scenes in the Hospitals	218
The Babies Taken and the Mothers Left	223
When Morning Came	224
The Death Roll	226
The Missing	232
Identified Dead in East St. Louis	237
Aftermath of the Cyclone	238
Insurance Companies Losses	242
Indirect Losses	244
Suffering Mechanics	248
Telegraph Companies Overworked	249
Two Days After the Calamity	255
Visiting the Cyclone District	259
Another Disaster Narrowly Averted	262
One Hundred and Forty Thousand, Actual Count	266
Half a Million Spectators	267
Feeding the Hungry	271
After the Crowds Had Gone	271
Along Broadway	273

CONTENTS.

	PAGE.
Burying the Dead	277
Sad Funeral Scenes	280
An East St. Louis Funeral	284
A Thieves' Opportunity	286

INCIDENTS OF THE CYCLONE.

Narrow Escapes, Deeds of Heroism, Records of Disaster and Desolation	289
Fate of a Harding, Ill., Merchant	291
Swam for His Life	292
Was Dug Out Unhurt	295
Dug His Own Way Out	296
Tried to Save Her Diamonds	296
A Perilous Voyage	297
Drifted to the Barracks	298
Wheels in the Air	302
Saved by a "Dumb Waiter"	303
Mysterious Munn Family	304
The Crime of Thieves	304
Heroic Work Done by Ambulance Drivers	307
Horse and a Live Wire Tangled	308
True Brotherly Love	309
How did it get the Dial	310
Chief Gross was Rolled	310
He Saved a City	313
A Janitor's Fate	315
Penned up for several hours in a Basement	316
Alone in the World	319
She Loved Her Dog	320
Shanty Boats Escaped	321
A Family nearly Exterminated	321
Crazed by Grief	322
A Pig's Souvenir	322
Killed on His Way Home	325
Capt. M. McMahon's Experience	326
A Humorous Incident	328
Tale of Two Families	331

CONTENTS.

	PAGE.
Rescued by an Officer	332
Mrs. Bene's Experience	333
A Jockey on Ice	337
Contributed a $50 Suit	337
He'll Never Say Die	338
Plucky Girl	339
A Coincidence	339
Pinioned in Debris	340
A Barber's Story	344
Her Daughter's Ashes	349
A Man Lost His Reason	350
Battle of the Winds	351
Families Broken up	355
A Terrible Scatter	356
A Man and Woman Blown Away	356
Horse Rescued Alive	357
It's an Ill Wind	359
Typical Sadness	360
A Wild and Awful Ride	360
The Wind's Idiosyncracy	362
How the Bird's Fared	363
Some Curious Effects	364
A Poor Refuge	367
Crawled Under the Bridge Approach	367
Alone with her Baby	368
Mrs. Worheide's Experience	368
An Adventurous and Costly Ride	369
Broke Up a Wedding	371
Eighty-Five Miles	372
A Ride on the Wind	373
A Kindly Deed	374
Another Lucky Escape	375
The Chinese and the Storm	376
Henry Wedermeyer's Experience	376
Uneasiness of Shanty Boat Owners	377
Strange Freaks of the Wind	379

CONTENTS.

	PAGE.
A Trunk Mystery	380
Havlin's Theatre Damaged	381
Engine House Completely Demolished	381
Killed Almost Opposite the Morgue	383
Panic Among Laundry Girls	383
A Priest's Escape	385
Died in their Arms	386
Hero of Balaklava Dead	388
Found Safety in a Basement	388
How the Horses Behaved	389
Mrs. Eyerman's Miraculous Escape	391
Gave His Life to Save a Horse	392
"Saved through God's Mercy"	393
Strange But True	395
Horse Unharnessed	396
Pet Cat Found	397
Exchange Members Timid	399
A Flying Stool and a Doll House	400
Three Little Ones Died	400
Blown into the River	401
Paper Hanger's Experience	402
Saved his Drink	403
Wedged in a Car Roof	403
Died from Fright	404
In a Fire-Proof Vault	404
Two Women's Escape	405
Sir Charles Gibson's Trees	405
Colored Victims	407
Willie Winckler's Death	407
Acts of Heroism	408
Saved his Captain's Son	409
Cabby was an Autocrat	411
Dying in a Drug Store	412
A Panic Averted	413
Train's Narrow Escape	415
Found a Dead Baby	416

LIST OF ILLUSTRATIONS.

	PAGE.
Entrance to Lafayette Park	23
Lafayette Park Presbyterian Church	24
General View in Lafayette Park after the Storm	29
Washington Statue, Lafayette Park	30
View from Relay Depot, East St. Louis	35
Wreck of Dr. Eyerman's Residence and Rear Portions of Union Club	36
Residence of August Ahrens, Whittemore Place	41
View on Mississippi Avenue	42
Residence of August Nasse, near Lafayette Park	47
Residence on Compton Heights	48
Leonori's Storage House, South Jefferson Avenue.	53
Portion of City Hospital	54
Front View of People's R. R. Power House on Park Avenue	59
Power house of Union Depot R. R. Co., on Geyer Avenue	60
Residence of Henry Roeder, Ann, near California Avenue	65
Lafayette Park Methodist Church	66
Side View of Lafayette Park Methodist Church	71
Purina Mills, Twelfth and Gratiot Streets	72
The Benton Statue, Lafayette Park	77
Lafayette Avenue, Looking West from Missouri Avenue	78
Lafayette Avenue, Looking East from Jefferson Avenue	83
Wreck of the Duestrow House	84
Scene on Lafayette Avenue	89
East St. Louis Transfer Co	90
St. Louis German Evangelical Church	95
Kœrner's Garden	96
Anchor Hall	101
Residence of Dr. Starkloff	102
Residence of Dr. Hauck	107
Scene on Mississippi Avenue	108

LIST OF ILLUSTRATIONS.

	PAGE.
General View of Lafayette Park	113
Residence on Park Avenue	114
Excelsior Laundry	119
St. Louis Jail	120
Wreck of Steamer "Henry Sackman"	125
Ottenad Furniture Co	126
Seventh and Rutger Streets	131
Russell Avenue, East of Oregon Avenue	132
Mt. Calvary Episcopal Church	137
Waverly Place	138
St. Louis Cotton Compress Warehouse	143
Gundlach Three-story Tenement House	144
East St. Louis Gas Works	149
Ploehn's Furniture Factory	150
Elevator at Chouteau Avenue and Levee	155
East St. Louis Electric Power House	156
Douglas School, East St. Louis	161
Residence of Dr. E. Preetorius	162
National Hotel, East St. Louis	167
Court House, East St. Louis	168
Agricultural Warehouse on Carroll Street	173
Mt. Calvary Church	174
Front View of Anchor Hall	179
Side View of Hodgen School	180
D. M. Osborne & Co.'s Building	185
Residence on Park Avenue	186
Wrecked Home on Oregon Avenue	191
Sample of the Storm's Rage on Park Avenue	192
Interior View of Ottenad Furniture Co.'s Building	197
Wrecked Trees in Lafayette Park	198
M. M. Buck & Co.'s Warehouse	203
View on Lafayette Avenue	204
Church on Mississippi Avenue	209
Collins' Livery Stable	210
Mauchenheimer Place, Seventh and Rutger Streets	215

LIST OF ILLUSTRATIONS.

	PAGE.
Residence of John Endres	216
Scene on Lafayette Avenue	221
Overturned Cars on Eighth and Park Avenue	222
A Ruined Home	227
Eads Bridge, East Side	228
Residence on Compton Heights	233
Wrecked Steamer	234
Desolated Homes	239
Union Club	240
Albany Dancing Academy	245
On Third Street	246
St. John of Nepomuk Church	251
Wreck of an Elevator	252
Freak of the Storm on Clifton Heights	257
Scene on Jefferson Avenue	258
Steamer "Exporter"	263
A Dreary Waste	264
Music Stand, Lafayette Park, Before the Storm	269
Music Stand, Lafayette Park, After the Storm	270
Hopeless Ruins	275
A Fallen Giant in the Park	276
Front View of City Hospital	281
Alone in the Ruins	282
Memorial Home	287
View of Schnaider's Garden	288
Ann and California Avenues	293
Century Building	294
View Across From City Hospital	299
Bridge and Wreck of Martell House, East St. Louis	300
Geyer and Ohio Avenues	305
Laclede Gas Company's Works	306
Hodgen School	311
Headquarters Relief Committee	312
Baptist Church, Lafayette and Mississippi Avenues	317
Dolman and Hickory Streets	318

LIST OF ILLUSTRATIONS.

	PAGE.
Texas and Allen Avenues	323
Where Children Dwelt	324
The Summer House in Lafayette Park	327
Unity Church	329
View at Twelfth and Gratiot Streets	330
A Sad Home-Coming	335
A Picturesque Ruin	336
Republican Convention Hall	341
A Promiscuous Wreck	342
Outside View Liggett & Myers Tobacco Factory	347
Second View Liggett & Myers Tobacco Factory	348
Liggett & Myers New Tobacco Factory	353
Trinity Church	354
Electric Car Blown from the Bridge	358
Scene on Chouteau Avenue	362
Union Dairy Company	366
East of Lafayette Park	370
Advance Elevator "B"	374
Wreck of Elevator on the Levee	378
All that was Left of a Beautiful Home	382
Imperial Laundry	386
Fourteenth and Papin Streets	390
Brown Tobacco Company's Building	394
Tri-Angle Warehouse	398
Wrecked and Blistered	402
St. Louis Wire Mills	406
Coming Home After the Storm	410
Singular Freak of the Wind	414

Foreword

Our world of news and information prides itself on quick response. Television crews report live via satellite from faraway places. Newspapers and computer services file by high-speed wire, and book companies flood stores with quickie paperbacks within days of a headline-grabbing event.

In 1896, the most reliable form of long-distance communication was the telegraph. St. Louis had had a telephone exchange since 1878 and was connected to Kansas City and New York City by long-distance telephone in 1896, the year of the Great Cyclone. Still, almost all city-to-city communication went via telegraph, a system that had become increasingly speedy and sophisticated in volume and delivery. The telegraph made it easy for city newspapers to print same-day news bulletins from across the country, sometimes even from places around the world. If the news was big enough, newspapers rushed onto the streets with "extras," and people rushed to buy them.

It was that growing modern appetite for immediate news that inspired the book, *The Great Cyclone at St. Louis and East St. Louis,* which is reprinted in this volume by Southern Illinois University Press. The original was produced with a speed not very far behind contemporary standards.

The Cyclone Publishing Company obtained the book's copyright in Washington D.C. on June 5, 1896, only nine days after the tornado churned like a turbine through the two cities, killing 137 people in St. Louis and 118 in East St. Louis. The "Publishers' Notice" in the first printing of the book is dated June 10, 1896. And although the first date of sale is not available, there is enough evidence in the text itself to show that the compilers were in a hurry.

The subject was worth the effort. The tornado that struck shortly after 5 P.M. on Wednesday, May 27, 1896, remains the single deadliest incident to befall the St. Louis area. Cholera epidemics killed many more people in the years before the Civil War, and the withering heat wave of 1936 accounted for 421 deaths. But those plagues recorded their tolls over time; the tornado did its damage in about twenty minutes. It blasted a ten-mile-long path through the Near South Side of St. Louis, roughly along the path of Interstate 44, across the Mississippi River at the downtown levee, and into the railyards and commercial districts of East St. Louis. It destroyed 311 buildings, heavily damaged 7,200 others, and significantly harmed 1,300 more. The loss, including destruction of riverboats, was estimated at $10 million to $12 million. (Because the federal Consumer Price Index had not yet been developed, there is no sure way to put that loss into contemporary terms. But to estimate, consider that in 1896, a tidy, two-story brick home in a good neighborhood could be bought for $1,500.)

Another motivation for the publishers of *The Great Cyclone* was that St. Louis considered itself worthy of the notice. The city was fast becoming the nation's fourth largest, behind only New York, Chicago, and Philadelphia. From 1890 to 1900, a decade held back by prolonged depression, the city grew from 451,770 residents to 575,238, a 27 percent increase. Thousands of home bookshelves included a book called *St. Louis, the Future Great City of the World*. Union Station, monument to the city's place as the nation's second busiest railyard, had opened in 1894. City Hall was not finished, but was open for business. Civic leaders had begun talking about what would become the city's most shining moment, its Louisiana Exposition, or St. Louis World's Fair, in 1904.

In 1896, most of the city's residents lived within three miles of the downtown river landing. The elite had left the Lafayette Park neighborhood southwest of downtown for the new tony address, Vandeventer Place, just north of St. Louis University in what is now Midtown, three miles west of the Mississippi River. The poor lived in small houses and tenements jammed around the edges of downtown and the industries along the river and railroads. Tidy middle-class districts were spreading past Grand Boulevard. Almost everything was built of the city's trademark ruddy-red brick.

The storm struck after three weeks of violent storms that swept across the country from Nebraska and Texas to New Jersey. For most of April and May, temperatures and humidity were well above normal. On the morning of May 27, the newspapers reported that terrible thunderstorms the day before had flattened crops in Missouri's Bootheel and swamped a ferryboat at Cairo, Illinois, drowning eleven. They also printed the daily forecast, "Partly cloudy weather continues, favorable for local thunderstorms."

The tornado that became known to St. Louisans as the Great Cyclone was the deadliest of a series of tornadoes that struck from central Missouri into southern Illinois on May 27. All told, 306 people died, according to the official account of the Weather Bureau office in St. Louis. The total included six killed in Audrain County, Missouri, about one hundred miles west of St. Louis, and thirteen more in New Baden, Illinois, thirty miles to the southeast. In St. Louis County, one child died when a tornado struck a farmhouse south of Clayton.

News accounts of the time and this book refer to the tornado as a "cyclone." Actually, a cyclone is any cyclonic wind movement, while a tornado is specifically a fast-twisting funnel cloud that can cause great damage. Thus, not all cyclones are tornadoes.

The tornado described herein touched down near the current site of the St. Louis State Hospital on Arsenal Street in southwest St. Louis, the highest point in the city. In 1896, the city's poorhouse, women's hospital, old folks' infirmary and insane asylum were located on that ridge. The city limit was further west near the River Des Peres, but this cluster of hospitals was on the outer edge of urban growth. The storm tore away roofs and porches and knocked out a few walls, but only eight people in this area were hurt.

That luck ran out fast. After striking in the southwest, the tornado then raced across Shaw's Garden (the Missouri Botanical Garden today) and struck just south of Tower Grove and Vandeventer Avenues, where the Liggett & Myers Tobacco Company, then a St. Louis-based business, was erecting a fifteen-building complex. Ironworkers were still high atop the four stories of girders when the tornado hit, and many were buried as steel and stout wooden supports collapsed.

The tornado then headed uphill toward Compton Heights, a district of prosperous families and nouveau riche not allowed into Vandeventer Place. Here, most every house at least lost some roof. Some neighborhoods, especially east toward Jefferson Avenue, were shattered. Whole blocks of homes lost their second floors, and rows of apartment buildings resembled honeycombs.

Reaching Jefferson Avenue, the storm crashed into what had become a busy gateway to the city's growing South Side. The bustle may have chased many of the rich westward, but the neighborhood around Lafayette Park was still upscale, and the park was still the city's pride. Prosperous clubs and businesses lined Jefferson, south of Chouteau Avenue. Two of the city's cable car systems, which moved cars along underground cables fed by giant steam-powered spools, had their power plants and shops there.

Much of this area was laid waste. The car companies were crippled, all but a few of the park's tall trees were snapped at their trunks, churches along the park were destroyed, and City Hospital, just east of the park, suffered heavy damage. Surprisingly, even though four hundred patients were housed there, only one was killed outright. Two others died a few hours later, including Franscesca Rodriguez who was reported to have died of "fright in the night."

But the worst carnage was further to the east, on the downslope of what has become known as the Soulard neighborhood, just south of downtown. Middle-class families lived on the ridgetop along Twelfth Street, income levels generally descending with the slope toward the riverfront industries. The corner of Seventh and Rutger streets, now a dingy parking area beneath the Interstate 55 viaduct south of downtown, was the deadliest spot—the "vortex," as the newspapers called it then. There, fourteen people died as the three-story Mauchenheimer tenement collapsed into a low, jagged pile of brick, timbers, and powdered plaster. Among the victims were Fred and Kate Mauchenheimer, who ran a busy tavern on the ground floor, and seven-year-old Ida Howell, who died in the arms of her mother, Alice, next to her father, a laborer named John. The discovery of the little girl's body the day after the storm was described with much sympathy in the newspapers.

Although the storm was strong enough to yank roofs from downtown buildings and smash skylights, the tornado's actual path curved just south of downtown. It crossed the river along a line from the St. Louis approach of the present-day MacArthur Bridge to the East St. Louis approach of the Eads Bridge, then the only dry way across the river downtown. Most of the sixteen boats in the harbor that day were wrecked.

The tornado knocked away part of the Eads approach in Illinois and blasted through the railyards and warehouses, which lined the East St. Louis landing and fed the still-busy Wiggins Ferry Company, a local monopoly. The tornado blew apart three locomotive roundhouses, the main riverfront elevator, and four freight stations, including the depot of the old St. Louis, Vandalia & Terre Haute Railroad. There, fifteen of the thirty-five depot workers were killed. Railroad hotels were also shattered. Beneath the ruins of one, the Martell House, a domestic worker named Mary Mock survived for two days and was rescued by diggers.

People who know St. Louis will recognize most of the locations and addresses noted in *The Great Cyclone*, but one is worthy of explanation here. The book, and the newspapers in those days, described the scene of the worst damage in East St. Louis as "the island." There was no effort to describe that name, just as news accounts today might mention St. Louis' Central West End without elaboration. The old island district is now the largely empty bottomland between the Illinois shore and the interstate highway as it curves past downtown East St. Louis. When a wide channel of the Mississippi ran along the path of the interstate before the Civil War, the land in between the two channels was known as "Bloody Island." There, St. Louisans such as Thomas Hart Benton settled grudges with duels. The Illinois channel silted in, but the name stuck into the twentieth century.

The "Publishers' Notice" of *The Great Cyclone* explains that its text was drawn almost exclusively from daily copies of the city's major newspapers, most notably the old *St. Louis Republic*. (The book's compilers are not identified, but their glowing praise of the "superb descriptive composition" in the *Republic* provides a good guess about where most of them worked.) Readers of the microfilm copies of the *Republic*, the old *Globe-Democrat*, the *Post-Dispatch,* and other St. Louis newspapers of the time will recognize almost all of the people and stories described in the book.

The photographs were not from the newspapers. During the time of the tornado, newspapers still printed engravings of drawings made by staff artists or artists hired for special assignments. High-quality artistry was of great value to newspapers in those days, and many of the scenes shown by the photographs in this book were faithfully rendered in newspaper drawings the day after the storm.

Although newspaper printshops were not yet ready to reproduce photographs in 1896, the photography of the day was of high quality—as long as the subjects stayed put. Beginning on the morning after the storm, photographers from studios in St. Louis, Chicago, and other cities swarmed through the wrecked neighborhoods. Several studios published "quickie" books of their own, most of them paperback booklets filled with clear photographs and texts of sometimes dubious accuracy. At least one version was printed in German, a language still much in use at the time within St. Louis's vibrant German immigrant circles. Newspaper accounts and photographs came together in *The Great Cyclone*.

The book's format may prove confusing. Short tales of individual incidents are scattered throughout the book, often without a flow of chronology or geography. A haste to publication may account for much of the disorganization, but that was also the way newspapers of the day provided their stories. Front pages of multi-decked screaming headlines would be followed by inside pages with stories arranged in no particular order of importance. One reason was the time-consuming and cumbersome nature of typesetting. Another was simply the confusion of newsgathering in a crippled city—reporters who could not use the crosstown telegraph or streetcars had to walk back to the office, where page nine may have been the only place left for an otherwise front-page story.

The writing style was also true to its age. The book's overview chapter, titled "The Great Cyclone," goes on at length about the "Storm King" and the "Fire King." There are numerous references throughout the book to lightning and funnel clouds as "fiends," and to the land as "terra firma." On page 105, the writers say of the wreckage at the poorhouse: "The situation was sufficiently horrible to unman the hardiest." The newspapers of the day read like that, too.

The book also includes a few inconsistencies that hasty compilation is prone to produce. For example, Francisca Rodriguez, the woman who died of fright at City Hospital, is identified as Francisco on page 97; six pages later, her first name is printed correctly. The book includes conflicting reports on the death toll at the Mauchenheimer house, as did the daily press in the days after the tornado.

Sadly, little information is readily available about the people who

FOREWORD

put the original book together. Business records no longer exist on Cyclone Publishing Company, which obtained the copyright. Nor is there ready information about Julian Curzon, the author and compiler who is described herein as "one of the most brilliant and popular magazine writers of the day."

But the volume is of great value as a journey along the tornado's path and as a glimpse into how the people of St. Louis in 1896 saw themselves and their city. It describes what people saw on the day of the storm and the few days afterward. It identifies many of the dead by name and describes how they died. It quotes from people who survived. It identifies businesses and homes by their addresses, helping readers a hundred years later to understand the cultural geography of St. Louis in 1896. *The Great Cyclone* is a thorough work of reporting from brick-strewn streets.

My thanks to the staffs of the St. Louis Main Library, the Missouri Historical Society, and the Mercantile Library in downtown St. Louis for their help in preparing both this foreword and an article on the one hundredth anniversary of the Great Cyclone for the *St. Louis Post-Dispatch*. My thanks also to Bob Broeg, sports editor emeritus of the *Post-Dispatch*, who was lucky enough to own an old copy of the book and kind enough to let me borrow it.

Tim O'Neil

The Great Cyclone at St. Louis and East St. Louis, May 27, 1896

PUBLISHERS' NOTICE.

SEVERAL business men, prominently connected with the relief movement in behalf of the cyclone sufferers, suggested that a book ought to be prepared giving a full history of the storm and its devastations, with a compilation of the many tragic and remarkable incidents connected therewith. Such a book, it was believed, would have a large circulation in all parts of the country, and supply a want that was manifest in the eager demand for news about the great tornado. It was also proposed that a certain liberal percentage of the receipts from the sales of the book should be donated to the relief fund, thus extending material aid in that direction while satisfying the desire for an authentic and permanent record of the most destructive and frightful calamity of the century.

This volume is the result of the suggestion referred to. Ten per cent. of the gross receipts will be assigned to the relief fund, to be used either for immediate necessities, or employed in assisting those who have lost their all in the destruction of their little homes, to re-establish themselves and make a new start in life, as the committees having these matters in charge may deem advisable.

The *nom de plume* of "Julian Curzon," as editor and compiler, will be recognized as that of one of the most brilliant and popular magazine writers of the day, and his connection with the work is a sufficient guarantee of its accuracy and literary excellence. Mr. Curzon, however, claims no credit for the splendid descriptive matter that occurs in these pages; this in a very large degree is due to the brilliant staff writers of the local press, whose articles have been collated and used as the basis of this work. But Mr. Curzon's active work in aid of the sufferers, his own personal experience and losses during the cyclone, and his connection with the various relief committees, made him acquainted with many singular and tragic incidents that have not been published elsewhere, and these are treated in his own brilliant and fascinating style.

The main description of the cyclone, its fury and its terror-inspiring horrors, is largely composed of that splendid and wonderfully brilliant report that appeared in "The Republic" the second day after the storm, which has attracted world-wide notice and comment as one of the finest examples of superb descriptive composition that has ever appeared in print.

The publishers of this work also desire to give due credit to the "Globe-Democrat," the "Post-Dispatch," the "Star," and the "Chronicle" for the use of material selected from their columns. In no other city could the press have displayed more energy, good taste, or literary ability in the handling of such an overwhelming calamity than was manifested by the great newspapers of St. Louis.

Our copyright is intended to protect the title and general form of the book, as well as a number of special photographic views that were taken for this work exclusively. The excellence of the views in general is due to the superiority of the workmanship and artistic conception of the local photographers, from whom they were purchased. They were all taken within one or two days after the storm, and represent the ruins as they then appeared.

THE PUBLISHERS.

ST. LOUIS, June 10, 1896.

INTRODUCTORY.

WE shall never know the full extent of the suffering caused by the tornado which devastated a section of St. Louis, Missouri, and a still larger portion of East St. Louis, Illinois, on the afternoon of Wednesday, May 27, 1896. It is known that more than three hundred people were killed and more than a thousand injured. It is known that a great portion of the city of East St. Louis was razed to the ground, and that South of and along the Mill Creek Valley in St. Louis, the cyclone cut for itself a wide path through block after block of residence property.

But it is difficult to even approximate the property loss and an accurate statement will never be made. The first estimates placed the total at $50,000,000. These figures have since been scaled down, but the total remains appalling. No estimate of loss can include the individual suffering, or the deprivation endured in silence by those, who too proud to ask relief, sought such shelter as was available and formed secret and praiseworthy resolutions to begin life over again.

In but a few minutes the savings of a lifetime were, in many instances, scattered to the four winds of heaven. Many were thankful to escape with their lives, absolutely penniless. There was not even time for them to secure possession of their pocketbooks and little keepsakes and mementoes. Death, destruction and desolation went hand in hand, and together brought about a reign of sadness and mourning, such as modern or ancient history but rarely records.

INTRODUCTORY.

One St. Louis citizen who was away from home at the time of the accident, stated on his return that but for the information imparted in the press, he never would have believed that the ruin could have been wrought except by the cannonade of an immense army, equipped with modern artillery. In describing his sensations on witnessing the scene of devastation he said that he was in Charleston just after the earthquake, and had entered more cities than one just after they had been stormed during the Civil War. But, he added, he had never seen destruction so complete or ruin so absolute as that wrought by the tornado, whose merciless devastation beggars description and calls for the use of words which would have to be coined for the occasion.

St. Louis is situated in the Mississippi Valley, on the western bank of the Father of Waters. A quarter of a century ago a cyclone blew through the neighborhood, causing great destruction in East St. Louis, but comparatively little on the Missouri shore. Since then there had been two or three trifling earthquake shocks. None of these had been sufficient to do any damage, nor had the thunder and wind storms which visited the city from time to time, wrought serious damage or caused general inconvenience.

The periodical floods in the Mississippi River, the last some four years ago, did great damage in East St. Louis, but practically none in St. Louis itself. In fact the people of the great metropolis of the Mississippi Valley States had for a quarter of a century been free from calamities of wind or water. The feeling of security had become general, and among the younger inhabitants particularly, it was thought that no cyclone or tornado was ever likely to penetrate the hills around the city and enter within its boundaries. The calamity at Sherman,

INTRODUCTORY. 19

Texas, had shocked every thinking man in St. Louis and a large fund had been raised for the relief of the sufferers in the Texas town. Many who subscribed liberally to the fund were themselves in need of assistance by the time the cyclone had reached and passed through their own city.

The awakening from this feeling of security was a rude one. The fatal day dawned with no exceptional occurrence. There was no friendly warning—there was no cry of "Flee from the wrath to come." True a cyclone had been unofficially predicted for the closing days of May, but the warning was not regarded, nor did those who were aware of it, dream that St. Louis itself would be smitten. Business was conducted as usual, nor was there anything in the condition of the weather early in the day to warrant any exceptional fear, or even thought. The weather bureau predicted local thunder storms, but said nothing of a cyclone, a tornado or even an exceptional wind. The sun shone as usual, but was frequently obstructed by clouds which towards noon became more numerous and threatening in appearance. The barometer began to fall with a steady persistency which alarmed those who have made a study of weather conditions, and who have learned what to expect from peculiar atmospheric conditions.

No one could tell the main direction of the wind, which seemed to come during the early afternoon in fits and starts from all points of the compass, veering around with sudden jerks. Towards three o'clock it became more settled from the Northwest with a number of sub-currents from different directions, which brought in masses of clouds. Gradually darkness seemed to approach and although the officials in the Weather Bureau Observatory do not seem even at this late period of the day to have anticipated a calamity, many people began to fear the worst. In one office building in particular the word was passed around that a cyclone was heading towards the city with lightning rapidity and that unless it was deflected from its course, a terrible calamity might be looked for.

Some received the warning as a jest, but others hurried to their homes and in some cases to their death. The office buildings of the city withstood the shock in a manner which redounds to the credit of their designers and constructors, although of course the full brunt of the storm did not strike them. It was the residence houses which for the most part were destroyed, and these were the most insecure places in which imaginary refuge could be sought.

At 4:30 it became obvious that the atmospheric conditions were unprecedented in the recollection of the people. The temperature fell rapidly and huge banks of black and greenish clouds were seen approaching the city. It gradually became darker and at 5 o'clock it was as dark in many parts of the city as is usually the case at the end of May, three hours later in the evening.

All the time the wind kept rising and in the far distance vivid forks of lightning could be seen. Gradually the thunder storm came nearer the city and the western portion was soon in the midst of a terrible storm. The wind's velocity was about thirty-seven miles an hour. This speedily increased to sixty, seventy and even eighty miles, by the time the storm was at its height. For thirteen minutes this frightful speed was maintained and the rain fell in ceaseless torrents, far into the sad and never-to-be-forgotten night.   J. C.

THE GREAT CYCLONE.

The Storm King and the Fire King combined in the attack on St. Louis. And right bravely did they fight. The air was filled with light and heavy debris, tributes to the might of the wind. In all directions the clang of the gongs of the fire engines and suddenly in the south there shot in the air a stream of flame that lit up the whole city. The Fire King had scored his first point in the assault. A big skylight came sailing past the observatory tower, circled around in the air, made a dive for Olive street and was stopped by the gutter along the building. Then down in the street could be seen the results of the attack of the invading hosts on the subtle agent of man, electricity. In every direction the long lines of telegraph poles were flashing pillars of blue flame. The wires were strings of fire and the insulators were blazing bunches of sizzling wires.

Buildings swayed and creaked in the powerful blast. The wind came down in the streets, picked up buggies and turned them over. It bounded to roofs, rolled up tin coverings like scrolls and deposited them in telegraph wires. It filled the air with flying bricks and timbers and made the ears horrified with the crash of falling signs and breaking windows and the shrieks of men and women. And all over the city fire engines hurried to and fro and flames broke out in such a multitude of places that the hearts of the brave firemen sank within them when they contemplated the possibilities of a general conflagration. But in this the allied forces of wind and fire worked at cross purposes. For the rain served to largely undo the work of the other elements.

It was 5:35 when the army of the air withdrew from the assault. Then the rain came down in torrents and drenched the throngs hurrying through the streets. It was a wonderful rain, a steady, pounding, penetrating rain that seemed to gather strength as it fell. Amid the horror and the wild rumors of countless fatalities the rain came down harder and stronger, gloomily sounding a knell.

A DISCRIMINATING STORM.

Had the arch-fiend himself directed the course of the tornado he could have scarcely guided it so as to have done more damage. Those who gazed upon the scene of desolation that was wrought were almost inclined to hint that it was guided by a strategist, so marvelous was its perspicacity. The forces at its command were handled with the skill and judgment a general of power might conceive. Carefully avoiding obstacles that might seriously scatter its concentrated power, it passed lightly over certain places in its path and smote others hip and thigh. Now and then it divided, but the divisions were strong in destructiveness, and where the main storm did not wreak its full force there was damage done in spots by wandering skirmishing blasts that appal and horrify. It switched and twisted and dodged with the skill of a

ENTRANCE TO LAFAYETTE PARK.

LAFAYETTE PARK PRESBYTERIAN CHURCH.

wrestler. And after inflicting damage sufficient to wear out any but the mightiest tempest, it still had strength enough to cross a great sheet of water like the Mississippi and spread ruin and desolation with a lavish hand. Though the tornado ravished the people of lives and property, they respect it for its power and ingenuity.

The rain and darkness of the night had the effect of bringing to the minds of those who worked in the afflicted districts the thought that experience shows what looks bad and terrible in the dark, generally turns out to be an exaggerated fancy in the morning. But the sun that rose smiling on St. Louis and her stricken sister city across the river the following morning, brought to light new horror with every ray that penetrated the atmosphere. It was not until the earth was bright and warm and the sky was clear and glad that the true situation dawned upon the people. The awful work of the devastating army of the elements stood out in all its naked hideousness. From the west to the east in the southern central portion of the town a wide streak was blazed, lined on each side with wreckage that represented to those who owned it when it was useful, a loss total and stunning.

THE CYCLONE'S FATAL PATH.

The path of the storm through the city was about seven miles long. It was not a direct path, leading straight from the point at which it effected entrance to where it left. It made a path like a snake striving to gain a place of known refuge from a pursuing enemy. Now and then it diverged from the Mill Creek Valley on one side or the other, but only to return at some vulnerable point with renewed energy. It seemed to move at a height above the ground that sheltered the low places in its path until after it passed the City Hospital. Then it came closer to the earth, and the damage wrought from Twelfth street to the river shows that it rushed directly down the incline to the Levee.

Where it first entered the city, out near the Poorhouse, on Arsenal street, the force of the storm was exerted against trees and scattered buildings. The first indication of the real force of the wind was made apparent at Jefferson and Geyer avenues, where the big power house of the Union Depot street railway system, one of the largest electric plants in the world, was razed, damage to be measured only by the hundred thousand dollars, being wrought. A block further north the destruction was, if possible, even more emphatically manifested in the wrecking of the Union Club building, and the almost total demolition of dozens of buildings in the immediate vicinity.

The force generated at this corner was not lost while the storm continued on its way east until after it left Lafayette Park, and there was no extraordinary manifestation until it reached the ill-fated corner of Seventh and Rutger streets. Here it spent the full vent of some of its reserve fury, and then moved on to Soulard Market, which formed another center of destruction, wider and longer than either of the others.

While the chief force of the storm was exerted at the three places mentioned, there were others scarcely more fortunate in the matter of locality in its path. Tower Grove Station, where the mammoth plant of the Liggett & Myers Tobacco Company was destroyed, was the first point where the surrounding circumstances warranted the wind in extending itself. When it reached the river it met a wide stretch of waste that allowed it to gather force and gave the scattered clouds a chance to rejoin the main body in the assault upon East St. Louis. But at no time was there any diminution in the vigor of such portions of the tornado as reached the earth. The scattering edges of it completed the waste the main body emphasized in spots.

ALL THE ELEMENTS OF A TORNADO.

There can be no doubt that the storm was a tornado rather than a cyclone. The local weather bureau observer, in explaining this point, says:

"It was at first believed that the storm was entirely a straight wind rush, without any evidences of the tornadic whirl. Further investigation late in the afternoon by one of our observers tends to confirm the opinion of many that the storm in a few circumscribed localities was a genuine 'twister.' This appears to have been the fact around Lafayette Park and on Chouteau avenue between Ninth and Twelfth streets. In these places the debris lay in every direction, and in the park trees had been torn up by the roots and thrown around in confusion, indicating the presence of the inward spiral and upward motion which is characteristic of the true tornado. This also accounts for the presence of the southwest wind while the general direction during the storm was from the northwest. In the tornadic whirl the wind would have an inward rotary motion, and as the center of disturbance progressed the wind would shift accordingly to the different points of the compass. It is probable that this southwest wind was of greater velocity than that from the northwest, according to one of the accepted general theories of cyclonic circulation, namely, that the heaviest wind may be expected in the southwest quadrant of a low area, and the heaviest rain in the southeast quadrant. No other evidences of the true tornado have thus far been obtained."

Apart from scientific observation there can be no doubt as to the tornado motion of the storm. Any man who looked at Lafayette Park on Thursday morning and then said there was not a rotary motion in the wind that tore down the kings of the forest, must have been afflicted with a disease that called for the use of a strabismometer. They fell to the north and to the south and to the east and to the west. If one had the time and the inclination—and perhaps the mind to figure it out—he could, while in Lafayette Park, have determined just about the number of times the storm cloud twisted in going across the inclosure.

But one did not have to go to Lafayette Park for proof that St. Louis was assailed by a storm with a rotary movement. Almost any cross street in its path furnished the evidence. On one street trees on the west side were blown down to the west and trees on the east side were blown to the east. On the next street east the

GENERAL VIEW IN LAFAYETTE PARK AFTER THE STORM.

WASHINGTON STATUE, LAFAYETTE PARK.

order was reversed. There are corners in South St. Louis on which buildings stood that give unmistakable evidence of having been struck by wind coming from more than one point of the compass at the same time.

WARNINGS OF DEADLY PERIL.

We have seen that although not officially predicted, the tornado was expected by many. The position of several of the dead bodies found in the wreckage indicated attempts to secure places of safety which must have been commenced long before the storm reached the city, or its suburbs. The cloud movements all the afternoon created alarm in the minds of those who had spent any length of time on the prairies in the West. The peculiar shades of green the vault of heaven assumed, denoted to the eye of the man who has gone through tornadoes where there are no compact communities to break them up, a state of affairs in the atmosphere calculated to provoke alarm. The scurrying clouds, sweeping about with apparently no object in view, were plainly bent upon a mission of destruction. On account of the popular fallacy that a tornado or a cyclone will not strike a large city there was created a feeling of false security. But men who know about those things were worried. Women who knew about them prudently made preparations to seek places of safety should the conditions warrant the passage of the storm over the great area of the town.

The gathering of the clouds in the west preparatory to the assault of the city was one of the grandest spectacles ever given the view of the eye of man, but it did not compare with the terrific grandeur of the onslaught. When the allied forces from the southwest and the northeast, laden with lightning and cumulative force, retired beyond the western horizon for a few minutes, shortly after 5 o'clock, Wednesday afternoon, and the rain came blowing in soft sheets from the south, it was thought that there would be no further disturbance. But the army of lightning and winds had gone for re-enforcements. Away out beyond the limits of the county there was another storm raging, a malignant storm that spent its force in doing damage of small moment in comparison with what it might do if joined to a greater combination. The whirling bank of clouds went out in St. Louis County and gathered up the angry elements warring there. It took about two minutes to make the consolidation and then came the combined attack.

GATHERING FORCE IN ITS COURSE.

The fearful combination of aerial forces which wrecked thousands of buildings and broke up as many homes, grew in strength as it dashed through the city in its death dealing course. The storm it had gathered in its ranks out in the high ground west of the city was not fully settled in position and confusion reigned in place of the discipline that predominated in a few minutes. In the center there was a thick mass of fighting strength, but the outside ranks were weak and wavering, prone to wander off with each fancy of the wind. Coming down toward the Poorhouse the rotary motion, the secret of the tornado's power, began to make itself apparent to a visible extent. The army of clouds, from the shape of a drawn out accordion, partook more of the shape of a screw, with the small end, the effective end, dangling towards the earth. Once the cloud was formed the onslaught was inevitable. The wind took up the course that drew the storm toward the city and the city was doomed. No human power could stay the rush.

In a tempest of discordant sound the tornado swooped down on the Poorhouse, worked its will and continued on its way. The Female Hospital, just across the street, was right in its path, but the vagaries of a tornado are hard to understand. It took a slight trend to the east and south, clipped off the roof of the west wing of the Female Hospital, and then with the big Insane Asylum looming up before it, full of possibilities for horrible execution, it shifted once more to the south and took up a path that brought it over the asylum farms.

Upon reaching King's Highway the tornado was in a fair way to last on a voyage through the city. The scattering clouds on the edges were rapidly closing in on the central mass, and the screw shape was becoming more pronounced. From the direction it was traveling it seemed, when it passed the Insane Asylum, to be bound for Carondelet, but the strategy and ingenuity that actuated its movements came into play and steered it to a path more productive of loss of property and life.

Koerner's Garden—the mecca of the cyclist, and of the convival generally—stands, or we should rather say, stood on the corner of King's highway and Arsenal street. When the tornado was two blocks away from it to the south one would have sworn it was on its way directly east. But it wheeled suddenly in its course and made a dash for Koerner's. The momentum it had attained toward the east was too strong to allow it to strike the garden direct, but it made a good, honest effort at demolition, and then went on its way to Tower Grove Park, directly to the northeast. It tore across Tower Grove Park in a diagonal direction, gathering strength as it traveled, leveling trees and uprooting plants and shrubs, and then struck out over the hill for the lovely dwellings on Compton Heights. Away to the east, down in the valley, the big buildings of the Liggett & Myers tobacco plant reared their heads to the sky, and the great steel girders seemed to attract the tornado as a magnet attracts steel. It paused a moment in the rush to Compton Heights, took on a stronger rotary motion and sped away for the tobacco buildings.

ON THE OUTSKIRTS OF THE STORM REGION.

But the effort of changing the course so suddenly had a rather demoralizing effect on the army of clouds. A detachment of the solid platoon went away on an expedition of its own down toward the breweries and the peaceful German homes surrounding them in far South St. Louis. The main body, scarcely affected by the loss of the deserter, continued on its way, and death and ruin a second later reigned in the big steel structures at the Tower Grove Crossing.

At this time the mass of clouds seemed to spread. The top of the mass flattened and swung from side to

VIEW FROM RELAY DEPOT, EAST ST. LOUIS.

WRECK OF DR. EYERMAN'S RESIDENCE, AND REAR PORTION OF UNION CLUB.

side with a motion that threw the small end about like a pendulum over the country in a most promiscuous manner. It bobbed up and down, now striking the earth and tearing a streak of waste in its path, now high above, the force it expended doing damage slighter than direct application of the rotary motion accomplished. The rain was falling in torrents and the sky was veiled in a muddy mist that cast a gloom on the earth more terrifying than the darkness of night.

From the point where the tornado started in the attack on Compton Heights the wonderful and peculiar movements it took became apparent. While the body of the cloud was hanging over the residence district the tail, the destructive end, was creating havoc half a mile to the north in the factory district along the Missouri Pacific and Iron Mountain tracks. Just before reaching Grand avenue the storm got together again, dropped down toward the earth and made a charge on the hill it had been approaching.

Trees went down before it, houses were unroofed, walls were blown this way and that, and it appeared that the entire population of Compton Hill would be blown clear over the reservoir. But the tornado took another track. Instead of breaking itself against the side of the incline leading west from Grand avenue it took a sudden bound in the air, passed clear over Grand avenue and the Reservoir Park and continued on its way over the valley between Jefferson and Grand avenues toward the more densely populated part of the city.

Though the storm did not directly strike here, it had

a few guerrilla clouds out that accomplished much in the way of destruction. These vagrant fighters against property and life skimmed along close to the earth, dodged around obstacles that were beyond their power to remove and tore and tugged at towers and roofs and trees. As they flew by, the main storm, mounting higher and higher, seemed to lose them, but they kept on, straight to the east. Meanwhile the tail was slowly swinging out to the north and the cloud flattened to a more pronounced degree.

THE FULL FORCE FELT.

Clear down to Michigan avenue the little skirmishing clouds did all the execution for the storm. Crossing this street the small end suddenly swooped downward, and at the same time the upper end began to swing around in such a way as to bring the cloud into an almost perpendicular position. At Ohio avenue the full force of the armament of the army of clouds was thrown at the earth for a second and buildings fell in every direction. The center of the storm was at this time directly over Lafayette avenue, and the tail again began to diverge to the north, with a graceful swing. The front of the advancing phalanx of deadly vapor was right on a line with the surface of Jefferson avenue.

Lafayette Park is on the summit of a hill that forms the western boundary of the valley extending up to the high ground along Grand avenue. Jefferson avenue is slightly down the incline to the west of the park. If the tornado had pursued the same antics that governed it when it reached Grand avenue, it would have again bounded in the air about Jefferson avenue and continued on its way toward the river at a different altitude to save from serious damage the beautiful residences about the park. But the actuating power showed strategy this time. There was not much to destroy at Grand avenue, but Jefferson avenue offered a rich harvest. And when the storm reached that thoroughfare it dashed straight into the side of the hill, the destructive tail swung east as far as Chouteau avenue, dashed toward the earth and bit a chunk of property out of Jefferson avenue all the way from Chouteau to Russell. The tail of the storm moved from north to south like the lash of a whip, and while it was completing the work of destruction the dense main body remained poised in the air, slowly revolving and floating in the direction of Geyer avenue.

About the time it was directly over the Scullin power house the tail came along, swept under, and with a roar that was heard for blocks, mixed motors, engines, cars, buildings, machinery and men in a mass of matter. Then the tail swung over to the South Side race track, completely licked it off the face of the earth, and the great body, flashing lightning and breathing thunder, moved swiftly to the northeast, blowing down houses and stripping Lafayette Park on the way, leaving it a forest of splintered stumps.

The movements of the tail of the storm—the twisting tail that curled and splintered iron and steel and melted away the strongest work of man, then moved so rapidly from one side to the other that it could not be distinguished. It slipped over to Twenty-first street, destroyed elevators, wire mills, warehouses and factories, and then considerately jumped over a big brewery at the south approach to the Eighteenth street bridge. It sent an arm over to Market street and blew holes in the walls of factories, but spared the Union Station. Then it gathered all its force, swept back to the south again, wiped the City Hospital out of existence and started on the voyage down Soulard street, the most destructive in its campaign. All the time the big black cloud hovered above, sliding up and down, spitting, lightning and raining.

WHERE DEATHS WERE MOST FREQUENT.

From the City Hospital to the corner of Seventh and Rutger streets the storm gathered itself together for a final onslaught before reaching the river. All the little tormentors were drawn into the vortex and they did not forget their work of destruction as they came from every side for blocks around. The terrible tail was drawn up into the main cloud as though for additional strength, and at Eighth street the entire mass dropped to the earth and spread dozens of whirling dealers of death to the north and south. Men and women and children died here like flies in the cold. The streets were blocked with debris, while the awful tail whipped up and down, and from side to side like something infuriated. The tornado tarried here, but not for long. It worked fast.

From the scene of carnage on Rutger and Soulard streets, from Broadway to Eighth, and as far north as

RESIDENCE OF AUGUST AHRENS, WITTEMORE PLACE.

VIEW ON MISSISSIPPI AVENUE.

Chouteau avenue and south as Shenandoah, the tornado gathered together again and went bounding toward the river. Part of it cut loose from the main body and started on a voyage to the north over the downtown business district, but the strong army was in the cloud bound for the last stand in the battle on this side of the river. All along the bank it twisted and tore elevators and warehouses, and those the wind did not reach lightning set fire to, so that nothing or next to nothing escaped.

The storm next reached the Mississippi river, made the levee and harbor a desolate waste, blew steamers and wharves to the other shore, and lashed the water to waves of proportions never seen in this vicinity before. Above the Eads bridge the squadron that broke away in South St. Louis rejoined the main body somewhere in the stream, and the united army went into East St. Louis, scarcely tired from the long flying battle that extended over eight miles of disputed territory in the Mound City and on the river.

THE ATTACK ON THE RIVER.

The first reports of the loss of life and injury to limb sustained by that portion of the tornado-swept district along the river front were, as was natural in the excitement of the moment, exaggerated. After a careful canvass of the situation, it was found that the loss of life was comparatively smaller on the river than in any other section visited by the storm. The officers of the various steamboat companies made careful inquiry into the matter, so far as the means at their command allowed, and nearly all of their employes and passengers were accounted for.

Although the broad bosom of the raging Mississippi offered a full sweep for the destructive powers of the tornado, it is a most remarkable thing that but an infinitesimal loss of life occurred in this portion of the storm-ravaged district, and only a few cases in which any injury was sustained by those whose occupation demands their presence on the river front. While the damage to the shipping in the harbor was something tremendous, it was regarded by river men as almost miraculous that the loss of life and limb is so inconsequential, especially in view of the accompanying destruction to property along the wharf.

While the storm was at its full height the air was laden with ponderous beams, iron rods, girders, planks and wreckage of every conceivable description, blown from the four quarters of the compass, and river craft of every class were crashing with destructive force against each other and grinding themselves into fragments, but few lives were claimed by the elements, and many reported dead turned up safe the next day.

There were sixteen steamers, five ferryboats, two transfer boats and six tugs blown from their moorings and completely demolished by the storm, besides six others which suffered comparatively little damage. The loss to shipping interests exceeded $1,000,000, as most of the boats have to be entirely rebuilt. Besides this, every wharfboat on the river front, except that of the Anchor Line, and those used by the city, was blown away and sunk in small fragments.

THE GREAT CYCLONE. 45

Among the boats which are a total loss are the City of Cairo, Arkansas City of the Anchor Line, the City of Vicksburg, J. J. Odil, Libbie Conger, Dolphin No. 2, which lies bottom up in the middle of the river; Bald Eagle, City of Quincy, Pittsburg, Belle of Calhoun, Harvester, Charles Merriam and George A. Madill.

The two latter are transfer boats. The steam launch Austria, said to be owned by Ellis Wainwright, which was moored at the foot of Carr street, was torn from her fastenings, lifted into the air and turned bottom up twenty feet from the shore.

THE RIVER A RAGING TORRENT.

The harbor boat, with Captain Jenks and five men aboard, was blown down the stream from Morgan street to a point below Lemp's Brewery. In drifting under the Eads Bridge her smokestacks went by the board with a crash, tearing out the front portion of the upper deck. Her upper deck aft was blown out on the trip down. David Welsh, a deck hand, who had sought refuge on this deck, was caught in the crash and pretty badly injured. The boat was towed back the next morning and anchored opposite the foot of Market street.

The Anchor Line steamer City of Monroe, which was loading at the foot of Pine street, was blown across the river and is now lying aground on the rocks on the Illinois shore opposite Chouteau avenue, minus her stacks, derricks, booms and a portion of her upper works. Had the storm held off a half hour longer she would have been cleared for New Orleans and escaped the storm. Fortunately no one aboard of her was injured.

The City of Cairo and the Arkansas City, which were lying just above the City of Monroe, were also blown out into the stream. They floated down to a point below the cotton compress plant near Carondelet, where they are now lying. There was no one aboard the two boats at the time except the watchmen, and they escaped injury, although the watchman on the Arkansas City was blown into the river opposite Market street. How he ever reached the shore in safety he is unable to say.

The City of Quincy, which turned bottom up in the river at Carondelet, was anchored at the foot of Franklin avenue when the blow struck her. On her way down she passed close to the harbor boat, and the watchman who was aboard of her clambered onto the harbor boat just before the Quincy capsized.

The City of Providence had a most providential escape from total demolition, and the dozen people aboard had a close call. The Providence and the City of Vicksburg were lying together when the tornado struck them. The wind blew through the cabin of the Vicksburg, which was closer to the shore, and caught the Providence, tearing her loose from her anchorage, blowing her out into the stream. The wind demolished the upper decks of the boat and ripped out the entire floor of the lower deck, scattering it into the river. After being blown to midstream the wind shifted and drove her back to the wharf, where she pounded a big hole in the Vicksburg's side, and tore out 100 feet of her own hull.

RESIDENCE OF AUGUST NASSE, NEAR LAFAYETTE PARK.

RESIDENCE ON COMPTON HEIGHTS.

Then the wind changed and blew both the Providence and Vicksburg across the river, the two boats pounding and grinding each other on the trip. The Vicksburg caught for a while on the Illinois side, and then was driven down below Carondelet, where she was demolished. The Providence was driven with great violence into the Illinois shore eight times on the trip down, causing great damage, and finally landed below the River des Peres on the other side. She was towed across to her landing the next day. The Vicksburg was a total loss, but the Providence was not so badly damaged.

ELEVEN LADIES IN DEADLY PERIL.

There was a scene of panic aboard the Diamond Jo steamer Pittsburg, which was lying just below the bridge, when the tornado struck her. There were eleven ladies aboard her, and when the wind began to create havoc aboard the steamer the terror was something awful. In their mad excitement the women attempted to leap overboard, and the captain, James P. Boland, and the crew had all they could do to prevent them from carrying out their mad intentions. In the midst of the excitement the wharfboat let go its anchorage, and began moving out into the stream with the Pittsburg, which was moored to it. Then an indescribable panic seized the passengers, and the captain, foreseeing the danger, yelled out for someone to cut the hawsers. This was a task fraught with the greatest danger to life and limb, as the air was dense with flying missiles, but Morris Killeen, the son of the superintendent of the Diamond Jo line, pluckily seized an ax, and, at the greatest personal risk, succeeded in severing the hawsers. The steamer parted company with the wharfboat, and was blown over to the Illinois side, but fortunately none of those aboard were injured. It landed opposite Carondelet, and went aground.

The Grand Republic, which was reported Wednesday night to be out in the storm with 500 excursionists aboard, returned to port the following morning without a scratch, the rumor having fortunately proven untrue. The Republic had gone up to Alton to take out an excursion, and was out of the direct path of the storm.

The Louis Houck from White river, and the Ferd Herold from Grand Tower, arrived in port on the 28th unscratched by the storm. The Spread Eagle from Alton and Grafton got in the same afternoon. The Idlewild from Cape Girardeau also tied up here on its regular trip. The tug Reliance towed up from Carondelet the wharfboat of the Eagle Packet Company, in a pretty badly damaged condition.

In its path across the Mississippi the tornado did not even spare the Eads Bridge, that monument to the engineering skill of one of the greatest men of the age.

The whole top abutment of the first pier, as well as the big rocks and iron girders of the approach, were picked up and thrown upon the roadbed just behind a passenger train. In front, two baggage cars were picked up and placed across the tracks. The top or upper roadway was torn down and thrown upon the train, as well as four wagons loaded with merchandise.

A sudden stop was made, and although the cars careened, the passengers escaped through the wreckage and were taken care of. All the way down the approach, every pole, signal wire and apron was torn away, and in many places piled on the tracks.

The hydraulic switch works shared the fate of all, and the pipes were scattered here and there along the approach and Broadway. The escape of the trains was remarkable.

THE DAMAGE IN EAST ST. LOUIS.

Nearly half of East St. Louis was wrecked. More than 100 people were killed and more than two million dollars of damage wrought on the east bank of the river. The damage was done in a few minutes' time, and how any person in the path of the cyclone escaped is a mystery to all who passed over the devastated section. The wind struck the levee just north of the East St. Louis elevator, about 5:30. The wharf of the Wiggins Ferry was the first to suffer, and it was thrown far up on the levee.

Eye-witnesses on the island graphically describe the approach of the destructive storm. Two of these men passed through the big cyclone of March 8, 1871, just twenty-five years ago, and they say the appearance of the clouds, the sky, etc., was similar to that of the former big wind. It struck the shore at exactly the same spot and passed up to town in the same way, never deviating more than fifty feet in either direction from start to finish. The cloud resembled an inverted funnel, and appeared to have the well-known and generally recognized rotary motion. When nearing the shore it seemingly divided and spread out, covering about 200 yards of a swath.

The scene from the east end of the Eads bridge resembled that of a battle-field. The dead and dying were removed from the ruins by willing workers, and the burning mills and warehouses lighted their funeral pyres with a distinctness that added horror to the awful scene. Lists of the dead were examined carefully to find some trace of relatives. At temporary police headquarters a special detail of policemen was appointed to keep the anxious inquirers out of the station. The officers stood about appalled by the devastation of the storm, and were unable to contribute anything to the record of the dead and injured. Hundreds of persons, however, told of the storm king's fury, and thanked God in the most devout manner that they had been spared to do so.

A JURY'S NARROW ESCAPE.

It was about 5:10 o'clock when the storm showed its strength in East St. Louis. The center of the tornado seemed to be between St. Clair avenue and the viaduct. Many dwelling houses in other parts of the city were demolished, but in this district there were few buildings that escaped. A law case was being tried by Judge Foulke in the court house. The jury had heard most of the testimony and, according to usual custom, were not supposed to be allowed to depart except in the custody of a bailiff. When the storm clouds began to gather. the foreman of the jury implored the judge to dis-

LEONORI'S STORAGE HOUSE, SOUTH JEFFERSON AVE.

PORTION OF CITY HOSPITAL.

charge them. The court house at that time was being shaken by the wind. Every moment its destruction was threatened. Judge Foulke at length decided to let the jurymen go. Five minutes after they departed for their homes that portion of the court house in which they had been sitting was torn away.

Timbers and bricks were piled on each other, and it is highly probable that had it not been for the forethought of Judge Foulke every juryman would have been killed. Judge Foulke himself succeeded in making his way to the Relay depot, where he was struck by a timber that was being blown through the air and badly injured. In the confusion that prevailed his body was allowed to lie on the platform for several minutes. People were too intent on ascertaining the fate of their own loved ones to stop to identify anybody, although it might be that of one renowned in legal circles and who had spent the most of his life sitting in judgment on his fellows.

At the hotels a panic prevailed. Women ran from one room to another, and along the corridors, screaming and seeking protection. Men who had faced almost every peril were powerless to comfort them. One glance from the window told them that the storm was one of greater force than any they had gone through or even contemplated. The strongest of them trembled and there were none who pretended they had reasonable hope of escaping alive.

It was all over in but a few minutes, but it seemed hours before the storm forced its way onward out into the fields and towards small cities and settlements, many of which were subjected to its ungovernable, inexplicable fury and murderous force.

PREDICTED A YEAR IN ADVANCE.

A prophet is not without honor, save in his own country. Hence it was perhaps that the warning issued in the summer of 1895 by Rev. Irl R. Hicks, the weather prophet, was not more generally noted and advantage taken of it. On the 17th of May, 1896, he repeated the warning. Few noted the fact, and fewer still gave it a second thought until the prediction was so dramatically and tragically verified.

Standing in his office and observatory two days after the storm this careful student of the planets and their storm influences, said:

"It was manifest for four hours before the storm broke loose, that we were in the vortex of a tornado. Everybody in this office was apprised of it. The readings of the barometer were an absolute guarantee.

"I stayed right at this window (a south window overlooking Twenty-second street) and pointed out to the gentleman behind me, Mr. Phillips, our advertising man, the exact location where the funnel cloud was forming, in which the tornado was organizing its vindictive powers.

"It was then about 3 o'clock.

"At the last moment when it was safe to be outdoors at all, just about half past four, after pointing out the exact quarters where the cyclone was organizing, I took the same gentleman out on the front steps with me and showed him the exact cloud formation in which the tornado had gathered its gangrenous forces.

"It was the last thing I did before ordering the porter to close the storm doors of the building. Having taken every observation possible for the safety of my office people and myself, I stepped back to the place already chosen in my mind as the safest in case the upper walls of the building gave way. I made all my people follow my example."

This place, as pointed out by the reverend gentleman, was the northwest corner of the large double parlor adjoining the sliding doors.

"The reason this place was chosen was that the frame of the folding doors would help support the mass of brick and debris that might pour down in case of accident. I think my action indicated whether I knew or not that the storm was tornadic in character and bulging with destructiveness.

"Now, it never has been mentioned that the very moment the tornado trunk passed east of a line on Twenty-second street there was immediately an illuminated shaft reaching up to the zenith, marking the exact location of the tornado trunk. It was swathed in dust and clouds and murky darkness, but it was nevertheless a shaft of fire. If it had been night time it would have illuminated the sky. That was the center of the tornado.

BAROMETRICAL READINGS.

Now for the facts. Our barometer read:

4 p. m. 29.10.   4:50 p. m. 28.98   5 p. m. 28.95.

"At 5:07 p. m., about the time of the passage of the tornado, when it broke, the barometer read 28.95.

"This was the lowest point touched by our instruments, the nearest record immediately prior to the explosion.

"But to put it still more emphatically. From 10 o'clock on in the morning the barometer's motions were phenomenal. Together with the aspect of the heavens they were an absolute guarantee that we were in the throes of tornadic violence.

"All these facts bearing immediately on the passage of the tornado followed out the special warnings that I had been issuing from month to month for a year. I emphasized the fact that there are causes bearing upon the subject, that they might be studied and made intelligible to the people. Now, as to the tornado itself.

"The monster was central over Lafayette Park. The outer whirls of it were circular, coming from every corner of the horizon, and involving more or less the entire city. Directly over the park locality there was a chaos of motion that is indescribable. On the northern tangent of the circle the motions were easterly. The westerly tangent dealt southerly and southeasterly blows, and so all around the firmament with the grand climax of the whole over the park.

"All you have to do is to look at the trees of the park. I counted in one corner of the park six trees that had fallen in as many different directions. Where they would not give way to a whirl they were stripped up. Tornado funnel clouds have a rebounding force, that accumulates as it passes along. Here it dips down, losing some of its force, and that accounts for some of the lulls in the track of the disaster, only to pick itself up again at some dis-

FRONT VIEW OF PEOPLE'S R. R. POWER HOUSE ON PARK AVE.

POWER HOUSE OF UNION DEPOT R. R. CO., ON GEYER AVE.

tant point for a renewed onslaught. The entire track of the tornado is marked by these bounding and rebounding spells, and the consequent greater or lesser havoc is created.

"There is no use saying that this tornado was worse than any other. The difference lies only in what came in its way. All tornadoes are alike. Had this cyclone swept a prairie region with a village here and there, fields of budding crops, forests and the like, we would not have considered it much of a disturber. But it struck a big city with a vast expanse of resistive material. Hence the awful destruction."

THE TEXT OF THE FORECAST.

In his Almanac for '96 Rev. Mr. Hicks made the following forecast for May: "Saturn, earth and sun will be on a line with each other on May 5, or in other words, the earth will pass between Saturn and the sun on that date, and, in astronomical terms, Saturn will be in opposition. The 4th is the central day of the Vulcan storm period, as shown in diagram on next page, with moon nearing the equator from the south declination. The Venus period is also shown in the storm diagram to be nearing the center on the 13th. Moon reaches both the apogee and the equator on the 7th, as indicated by the letter A in both diagrams.

"All things considered, a storm period of considerable force will run from about the 2d to the 7th. A brief high barometer and cooler weather will come in the track of the storms, but a change will quickly follow as we enter the reactionary period to run from about 9th to 12th.

"This will be near the center of the Venus equinox and at new moon. The opposition of the planet Uranus also occurs on the 12th. It will be prudent to watch all storm approaches from about 10th to 13th, as dangerous storms, with tornadoes possible, are very probable at this time.

"Another Vulcan period is central on the 16th, covering the 14th to 18th, with Venus and Mercury both bearing upon it, and each near the center of its period. A series of very heavy storms, with enormous rain falls, hail and thunder, will occur for days in succession at this time. The moon's first quarter falling on the 20th, within a few hours of the equator at C., and upon the central day of Mercury's equinox, the disturbances are apt to be continued, in a great measure, into the reactionary disturbance, running to the end of Mercury's period and the moon's perigree on the 26th, the stars on the 21st, 22d and 23d indicating central days of a specially threatening, reactionary period. Remembering that the Venus period is characterized by extremes of hot and cold, it will be prudent to anticipate very low temperatures after all general storms, with danger of frosts in northerly directions, until after the full moon on the 26th.

"The last storm period is central on the 27th, covering 26th to 30th. At this time the moon will be at the southern extreme of her declination, passing to the front of the earth to reach her last quarter early in June. Watch the storm clouds during all the active disturbances in May, as the probabilities are great for many dangerous storms. The storms at this season usually come in cycles, or periods of twelve and twenty-four hours, for a number of days in succession. The indications are good for such results during this month, and that many storms with heavy Venus downpours of rain and hail will result from about new moon to the end of the month seems most probable."

"I have already noted," said Rev. Mr. Hicks, commenting on his predictions and their fulfillment, "that the barometer's motions were phenomenal all day. The sky was decked in mackerel clouds, the infallible precursors of a cyclone. At 3 o'clock in the afternoon there was no longer any question as to whether it would break loose over this city or pass over us. The certainty was absolute that we were 'in for it,' to use a vulgar term.

HOW LOSS COULD HAVE BEEN LESSENED.

"At 3 o'clock on the fatal and fated day telephone messages could have been sent to every fire engine-house in the city. These messages, to the effect that a tornado was coming, could have been authoritatively supported by statements from the signal service office. The ringing of the fire bells could have been ordered.

"The crew of the engine houses could have been instructed to tell the people.

"The curious and apprehensive would have flocked to them.

"They would have been apprised of the coming of the tornado, and even roughly counseled as to what to do.

"The comparatively small loss of life and the stories of miraculous rescues illustrate better than anything else that many of the people knew what was wise to do under the existing conditions. Even the panic of the sudden and unforseen that-ensued did not stop many from choosing the proper method for protection.

"Like the proverbial spread of bad news with seven-mile boots, the warning would have been carried in less than an hour all over the city. To be conservative, it would have reached the larger part of it, for the vortex was central.

"The result would have been that people in many cases would have fastened up their houses securely and gone to the cellars for safety.

"The large manufacturing concerns could have been notified by telephone of the impending disaster, that they might have dismissed their employes and shut up their plants.

"The business population would have reached home quickly and in safety.

"Men could have been with their families to succor and support them in their fright, and hours, nay days, as facts have proven, of anxiety and suspense could have been spared the separated members of a household.

"The transportation companies and the street railways could have put under shelter much of their rolling stock which became playthings of the storm. Animals could have been driven to a place of safety, the boats of the harbor could have sought refuge further up or down the river, for captains are not ignorant of a tornado's course. Altogether, though destruction was inevitable, it could have been minimized by proper precautions, instead of breaking in over a careless, happy-go-lucky people, that

RESIDENCE OF HENRY ROEDER, ANN, NEAR CALIFORNIA AVE.

LAFAYETTE PARK METHODIST CHURCH.

set much store in the old, cherished idea that St. Louis had always escaped when other points were down. "But, as Schiller says: 'An eternal compact cannot be made with fate!'"

DEATH IN THE WIND.

St. Louis might also have been alarmed at the fearful record of death and disaster from cyclones during May. Prior to the culminating horror at the great Western metropolis—equal in magnitude to all that preceded it—the following was the ghastly record of the cruel and heartless wind for the month:

DATE.	PLACE.	KILLED.	INJURED.
May 4	Fairmount, W. V	2	3
May 12	Elkhorn, Neb	..	1
May 12	Lincoln, Neb	..	5
May 12	Sterling, Kan	..	3
May 13	Marshall, Okla	..	1
May 15	Sherman, Tex	85	140
May 15	Justin, Tex	1	19
May 15	Gribbe Springs, Tex	4	15
May 15	Howe, Tex	8	9
May 15	Farmington, Tex	4	3
May 15	Carpenter Bluff, Tex	..	6
May 15	Newton, Kan	1	2
May 16	Reading, Pa	1	2
May 17	Elva, Ky	5	2
May 17	Sympsonia, Ky	..	2
May 17	Seneca, Kan	8	8
May 17	Sabetha, Kan	5	9
May 17	Morrill, Kan	4	..
May 17	Frankfort, Kan	..	8
May 17	Onedia, Kan	6	..
May 17	Reserve, Kan	5	21
May 18	Lamoni, Io	..	4
May 19	Falls City, Neb	4	12
May 20	Eldon, Mo	..	5

DATE.	PLACE.	KILLED.	INJURED.
May 20	Versailles, Mo.	1	..
May 21	Lyon County, Kan.	..	2
May 25	Polk County, Io.	11	6
May 25	Jasper County, Io.	12	12
May 25	Manchester, Io.	..	2
May 25	Monroe, Ill.	2	1
May 25	North McGregor, Io.	13	17
May 25	Elgin and Mount Morris, Ill.	2	7
May 25	La Cygne, Kan.	4	..
May 25	Groveland, Oakwood and other points in Michigan	39	110
May 26	Cairo, Ill.	11	..
	Totals	230	437

CYCLONES IN HISTORY.

The great St. Louis cyclone, while not absolutely the worst on record in the matter of death and destruction which is scattered around, stands among the very worst. Before describing the scenes of desolation and counting the dead and wounded, it is of mournful interest to look into history and briefly tabulate some of the storms which have smitten different countries and communities.

London, because of its great age and extent, has been hit often. A hurricane's assault upon it as far back as 944 is recorded, which destroyed 1,500 houses, but the loss of life as given by the different chroniclers varies widely, ranging from 1,000 to 6,000. This is the earliest authenticated account of a calamity of this sort in London. The town was struck again in 1091, and 500 houses were knocked down this time, and from 200 to 400 persons were killed. November 26, 1703, several thousand persons were drowned in floods along the Thames, in London, which was preceded and accompanied by a hurricane that destroyed property in that town valued at $10,000,000. This particular tempest was called the "great storm," and figured as a time mark for half a century or over in "locating" dates of births, marriages and other events in the lives of people of England, for it extended throughout most of the country. It did, in this respect, the same sort of duty that the "big wind" in Ireland did a century and a quarter later. October 28, 1838; July 11, 1874; April 11, 1878, and December 12, 1883, brought storms to London which destroyed from twenty to thirty lives in each case, and from $1,000,000 to $3,000,000 property.

Havana had a storm on October 25, 1768, which killed 1000 persons and destroyed 4048 houses. December 12, 1822, Dublin and its vicinity lost from forty to fifty lives and $1,000,000 of property. Dublin was struck again on the night of Janury 6 and the morning of January 7, 1839, by a hurricane which was felt throughout the greater part of Ireland and on the west coast of England. Several other large towns in Ireland besides Dublin were attacked, including Limerick, Galway and Athlone, and in Liverpool the destruction was also great. Thirty persons were killed by falling buildings or flying missiles in that city, and 100 were drowned in the storms. In the four Irish cities named the devastation was greater, the loss of life in Dublin alone being put at 100, and in property at $3,000,000.

A LANDMARK IN IRISH RECORDS.

This is the "big wind" referred to in a preceding paragraph. It forms as important a "landmark" in the social history of Ireland as Magna Charta or the Revolution of 1688 does in the political history of England, or as the Declaration of Independence or the fall of Fort Sumter does in that of the United States. It is a great date line in the lives of many of the Irish people, who are 50 years of age or upward. When they say they were born in the year of the "big wind," or one, two or any other number of years before or after that time, they have, they think, given their age with sufficient exactness for practical purposes, and in many cases this is as near as they can get at their exact age.

Lisbon had a gale December 13, 1864, which was memorable in the records of that city of manifold and miscellaneous physical afflictions, destroying 100 lives and $2,000,000 property. In Maceo, Hong Kong and other cities in India on Septembr 22, 1874, a typhoon killed 10,000 people and wrecked between 40,000 and 60,000 houses. Buda-Pesth had a storm June 26, 1875, which destroyed 200 lives. In Haifong and other Chinese cities 300,000 persons perished in a typhoon October 8, 1881. At Colon, Panama, and in its harbor 100 lives were lost December 2, 1885, in a gale. Madrid lost ninety lives in a storm May 12, 1886. In Dacca, India, April 7, 1888, 100 persons were killed in a gale. Mauritius, in the Indian Ocean, on April 29, 1892, had a hurricane which killed 150 persons.

SIDE VIEW OF LAFAYETTE METHODIST CHURCH.

PURINA MILLS, 12TH AND GRATIOT STREETS.

THE GREAT CYCLONE. 73

STORMS IN OUR OWN LAND.

In the United States, Louisville was the first important town visited by a destructive storm. August 27, 1854, it had twenty-five persons killed and $1,000,000 property destroyed in a cyclone. Indianola, Tex., was nearly wiped off the face of the earth September 15, 1875, losing 126 lives and $1,000,000 of property. McAlester, I. T., nearly equaled that record May 10, 1882, the destruction being 125 lives and $500,000 of property. The year 1882 had a series of cyclones which were not matched in number and destructiveness until 1896. In one of these the town of Grinnell, Io., was destroyed, with a loss of $1,000,000 in property and 100 lives. Just seven days later Emmettsburg, Io., was blotted off the map, with a loss approximately equal to that at Grinnell. Mount Vernon, Ill., on February 18, 1888, was almost destroyed, thirty-nine persons being killed and 125 injured, many of them fatally. Pittsburg lost fifty-three lives in a tornado January 9, 1889.

Now comes the most destructive storm which ever visited a city on the Atlantic seaboard in the United States. This was the great blizzard of March 11 and 12, 1888, which New York City felt in its fiercest form. For more than a day business in that city was entirely suspended, and communication between it and the rest of the country was cut off as effectually as it was between St. Louis and the outside world during the tempest of Wednesday night of this week. New York for about twenty-four hours got its news from Philadelphia by way of London. About 400 lives were lost in New York and the other towns afflicted and in wrecks on the ocean in the vicinity, which the gale caused, while the property loss was estimated at $3,000,000, but New York City's share of this destruction was only about thirty lives and $1,000,000 property. Roscoe Conkling, who refused to be quarantined down town by this storm, and who walked from his office to his hotel through the mountainous snow drifts, travel by car, coach, vehicle of any sort or on horseback being suspended and impossible, died a few days later from the effects of his fatigue and exposure.

Louisville now figures a second time in a tornado, and sustained more damage than any other town in the United States from a wind storm, except St. Louis, in the hurricane of May 27, 1896. This was on March 27, 1890. Louisville's loss of life was placed at 125 at the time, and its property loss at $2,500,000. Like the St. Louis storm, the Louisville gale cut a broad path through the city, its width, in the course of greatest destructiveness, being fully 1,000 feet. The greatest storms in the United States since that at Louisville were those on the South Atlantic coast on August 28, 1893, which committed havoc in Charleston and Savannah, and with the shipping on the ocean in their vicinity, the aggregate loss of life being about 1,000, and in the Gulf of Mexico, on the coast of Louisiana, on October 2, in that year, in which 2,000 lives were lost.

Several series of storms occurring in the past twenty years are omitted here because the records give the losses by counties and not by towns. The most notable and destructive were those which visited Missouri, Illinois,

Kentucky, Tennessee, Mississippi, Georgia, the two Carolinas and Virginia, on February 9, 1884, in which the aggregate loss was 800 lives and 10,000 houses. In the region covered there were sixty different tornadoes on that day. The loss of life in the aggregate was greater on that day than was ever known before or since in that length of time in wind storms in the United States. The destruction during the last two weeks of May, however, beginning with the cyclones in Sherman and other points in the Southwest and ending with those in St. Louis and vicinity, and including the waterspouts and other meteorological disturbances, broke all records in this country for any equal period.

VICTIMS OF THE CYCLONE.

An account of the evil wrought by the storm must read like the annals of a great battle or series of battles. We have seen how the storm was predicted, how the clouds gathered in formidable and threatening array, and how they advanced upon a city which considered itself practically safe. We must now note what the cyclone or tornado did in the way of killing and maiming and what astounding damage it wrought to property. Thousands followed the cyclone's path the following day, picking their way among the wreckage, and pausing to gaze horror-stricken at some of the scenes that met their gaze. There was little said by the spectators. Their silence was more eloquent than any words could have been. Strong men shivered with horror, and if there were any unmoved, their callousness was well hidden.

It was a stricken city, or rather section of a city, that they saw. The dead were being removed from beneath piles of debris; willing workers were clearing away wreckage, hoping against hope that those who were known to be lying below had miraculously escaped certain death. Shade trees snapped, windows broken, chimneys prostrated, sections of tin roof ripped, cornices shattered, weak walls cracked—such things could be seen almost anywhere from the water works to Carondelet. But apart from these general evidences, there was the path of a storm cloud covering the city at its widest part, with a course from southwest to northeast. The path was as distinct as the boundaries of blocks. Beside the damage wrought within this path, the injuries sustained by the city at large were insignificant.

SURVEYING THE WRECK.

It was no easy task to follow the course of the storm. Where the wind had rushed the preceding day at a speed seldom attained by the fleetest locomotive, the foot passenger followed at a snail's pace. He had to walk, for no vehicle could be driven through the mass of ruins. Progress was from side to side by a zigzag route. This tracing of the course of the storm cloud began in the southwestern part of the city at Shaw's Garden. It ended at the Eads bridge. The distance across the city from west to east, at the widest part, is six miles. As the

THE BENTON STATUE, LAFAYETTE PARK.

LAFAYETTE AVENUE, LOOKING WEST FROM MISSOURI AVENUE.

THE GREAT CYCLONE. 79

storm traveled from southwest to northeast, the length of its path within the city was eight miles.

West of Shaw's Garden the storm cloud gathered. It did little damage immediately west, but on the north it unroofed buildings at Forest Park Heights, and on the south it caused the collapse of two large buildings at the Poorhouse, and did miscellaneous damage, altogether amounting to perhaps $50,000. Singularly enough, the Insane Asylum and the Female Hospital, standing on the same elevation as the Poorhouse, escaped anything serious. Shaw's Garden, in the wide Valley between Forest Park Heights on the north and the group of city institutions on the ridge to the south, showed dismantlement. One of the largest conservatories was crushed in. Other buildings were injured. The trees and shrubbery were torn and scourged.

In its beginning the storm cloud, as we have already seen, started from several points. It showed its power at the localities high and low, mentioned, but in the territory between it left weaker buildings unharmed. And thus it moved along in strips up the easy grade toward Grand avenue, half a mile eastward. It was not difficult to trace the southwest section of the cloud from the Poorhouse hill across Tower Grove Park and to the Old People's Memorial Home, on Grand avenue, by prostrated trees, overturned fences and some injuries to houses. There was a like strip from Shaw's Garden eastward toward the Compton Hill reservoir on Grand avenue. A third division of the storm cloud, perhaps the one which unroofed buildings at Forest Park Heights, moved eastward on the north side of Shaw's Garden and tore down the steel frame work of two of the group of buildings being constructed for the Liggett & Myers tobacco factory. This steel frame work was four stories high. It towered above the unfinished brick walls three stories. Three big buildings on an east and west line stood at that stage of construction.

The section of the storm cloud passing north of Shaw's Garden tore down the three uncovered stories of steel beams and supports as if they had been so much of cob houses. This it did with the first and third of the buildings, leaving the middle building unharmed. It was under the mass of structural steel at the third building that eight lives were crushed out. A little beyond the Liggett & Myers plant the cloud lifted ten heavy freight cars from the St. Louis and San Francisco track and laid them on their sides. From that point this third section of the cloud made its way eastward.

On Grand avenue, between Lafayette on the north and Shenandoah on the south, the three sections converged and united their forces. Between the paths traveled by the sections of the cloud to the immediate vicinity of the reservoir stretches of the city sustained no loss whatever. Along the three paths were continuous trails. But after crossing to the east side there was but one path, wide and extensive.

A SUBURB DESOLATED.

Compton Heights is the name this suburb adjoining the reservoir bears. It is one of the highest points in St.

THE GREAT CYCLONE.   81

Louis, having been chosen for that reason as the water distributing location to the southern half of the city. A portion of the heights has been laid off in accordance with advanced landscape gardening ideas, with curving roadways and building sites of irregular form. Compton Heights, ten blocks from north to south and six to eight blocks wide, has 300 residences which cost from $5,000 to $25,000. Not one of them escaped some injury and the most of them show ugly gaps in their walls, demolished gables and missing sections of roofs. These are houses of recent construction, strongly made and well finished. Through the rents which the storm cloud made the torrents of rain poured in and drenched the interiors. The outward desolation of Compton Heights revealed but a part of the loss sustained. Richly upholstered furniture, books, tapestries and velvet carpets were scattered on the lawns to catch the drying sun and wind. There were streets on Compton Heights which were impassable until choppers went through them cutting and rolling out of the way the prostrate trees. Trees as large as a man's body were torn up by the roots and when the roots came up they broke and displaced the squares of the granitoid pavement.

The storm cloud passed from Compton Heights, bearing a little to the north of east. It struck a different class of houses and a different neighborhood. Moving up and lapping over to the north of Lafayette avenue, while it angled away from Shenandoah on the south, the cloud was a full half mile wide on Oregon avenue. From Oregon eastward to Jefferson avenue is four long blocks, dipping gently to Ohio and then rising again to Jefferson. In the forty blocks lying within these boundaries the rule is lots of twenty-five feet and substantial houses of from $3,000 to $6,000 cost. The territory is closely built. It contains over 500 houses. Fully 200 of them were left uninhabitable.

Three-story houses were reduced to two stories, and there was nothing but the blue vault of heaven above those who went to bed upstairs. Sidewalls from roof to foundation were cut out. Sometimes it was the west wall and sometimes it was the east wall. Bay windows were suddenly enlarged to treble the size planned by the architect. Roofs were gone. In thirty or forty cases the whole house was a heap of brick and timber. Near the corner of Russell and California avenues the storm took the roof off one house and moved the roof of the next house over upon the first house. It was a sad misfit, however, Talk about the architecturally weak being the sufferers! On California avenue north of Lafayette is an almost new public school building. The storm cloud took out a part of the sixteen-inch wall on the north side. There are livery stables, drug stores, groceries and markets in this district between Oregon on the west and Jeffereson on the east. They all suffered as did the residences. The damage was universal.

LAFAYETTE AVENUE, LOOKING EAST FROM JEFFERSON AVENUE.

WRECK OF THE DUESTROW HOUSE.

IN AN OLDER SECTION.

At Jefferson avenue the storm cloud struck the Lafayette park district. Its northern line edged over Park avenue. Its southern limit was the South Side race track. The path widened into three-quarters of a mile. The storm cloud came to one of the most complete of the smaller parks in the United States. It found the finished work of thirty years of municipal care. Lapping over the park two blocks to the northward and four blocks to the southward, the cloud made its deepest impress upon the city.

People who had wondered what would happen if a tornado should strike a great city, wondered no more after inspecting the stricken Lafayette park district. Thomas H. Benton in bronze stood on his pedestal, pointing to the west in the pose suggested by his words.

"There is the east, there is India." Around him was the park forestry of a third of a century's growth, elms, maples, poplars and fir trees. Benton's statue remained in place, but all around him lay trees prostrate, while those not torn up by the roots stand gaunt, twigless and leafless. The massive iron fence was flat on the ground. People walked in and out where they will. The gates were useless.

St. Louis' beauty spot is gone. Lafayette park, the queen of all the city's outing places, had been suddenly transformed into a vision of ugliness, whose imperfections now were accenuated by the recollection of its former comeliness. Lafayette park was a favorite with St. Louisans. Its destruction caused many a child to lose its cherished playground. Nurses will for months be at a loss to know where to take their charges for a breath of fresh air and many of their elders will sigh for the cool nooks and foliage for which this retreat was famous. Situated in the direct path of the storm center, nothing escaped the fury of the cyclone except the statues. The granite columns were proof even against a tornado and stood the day after the storm like grim sentinels guarding vandals from adding spoliation to devastation.

Thousands of persons visited the park and commiserated with each other on the ruin of what had been justly called the prettiest spot in St. Louis. They cut canes from the branches of the trees and carried away parts of the benches as mementoes of the storm. They saw the bark torn from the body of huge oaks; saplings were torn up by the roots, and for the first time they gained an adequate idea of the force of a quantity of compressed air driven by a power no one has analyzed. Part of the iron fence surrounding the lake was blown into the water. All the houses except that occupied by the caretaker of the park were destroyed. That building was damaged badly. The band stands were blown away and parts of them were found hundreds of yards away. Every flower in the park was destroyed. Some of the rose bushes were uprooted, while the smaller varieties were crushed under branches and flying debris. At every entrance of the park there were trees lying across each other. The only motive of the cyclone seemed to have been to typify fantastic

THE GREAT CYCLONE. 87

anarchy. Disorder prevailed, for in the madness of this storm there was no method. The only living things in all the park, except the sightseers, were the birds on the pond. Even they swam about in an aimless, bewildered sort of fashion.

MOURNING THE RUIN OF THE PARK.

It is a noteworthy fact that indescribable as was the destruction of homes, and heartrending as this harbinger of poverty was, the ruins of Lafayette Park excited the strongest and most universal expressions of sorrow. The residents of the South End in particular were distressed at its misfortune.

To them the park was a thing of beauty and a joy forever, and more. It was the scene of many tender memories. Were it not for the storm, many of them would have been rowing in the park pond to-day. Others—staid citizens and tired youths—would have been reading papers and books beneath its leafy bowers. The children would have been romping over its pretty lawns and the mothers and daughters would have been admiring the exquisite beauty of the flowers.

The young folks, as they picked their way through the network of tangled trees, recalled the many pleasant hours they passed there. They skated on the pond in winter, they boated there in summer, and when the snow was hard and glassy they coasted down the flagpole hill. As children they had innocent flirtations under the shade trees, and when they became older and more formal they visited the park on Thursdays in summer and listened to the music of the band, gossiping between the numbers. This is why the South Enders mourn the havoc played in the park almost as much as they grieved over their own desolation.

To none did this seem sadder than to Sir Charles Gibson. When he looked out from the wreck of his palatial home and cast his eyes over the once beautiful park, the tears coursed down the old man's cheeks.

It was not for his home that he cared most; not for the dollars that the storm's fury had stolen from him; not for the litter and the debris that littered the beautiful avenue. His own flesh and blood were safe—his dear ones—and next to them his heart went out to the poor, distorted trees which for forty years he had seen growing from tiny little shoots, like little babies, into the great massive trunks fit to grace the primeval forest.

This was desolation no gold could replace. This was a wound to him no time could heal. He is an old man now, and as he stood there contemplating the sad scene it all came to him. He could never again see that picture that had so often made him young and carried him back to the days when his blood was coursing strongly in his veins and life was roseate with the flush of manhood's prime.

"Dick," he said to his son-in-law, Richard Perry, who stood beside him, "it took forty years to grow those trees, and I shall never see their like again. The house I can repair, but my trees are gone forever."

The Lafayette Park churches, which had been gay with summer costumes and bright faces, will not echo to

SCENE ON LAFAYETTE AVE.

EAST ST. LOUIS TRANSFER CO.

hymns of praise for many days to come. The crumbling walls did not invite one to prayer the day after the storm, but even if they were intact, they would have few worshipers. The reason was that many families whose names were on the membership rolls were bereft of homes, and all too disconsolate to seek comfort by their usual devotions.

HOMES AND CHURCHES DESTROYED.

Fronting on the four sides of Lafayette Park were the five stone-front mansions, which were the pride of the city fifteen years ago, before the West End movement set in. Opening out from the broad avenues around the park were Benton, Nicholson and other "places," with the fringes of beautiful homes. Four stone churches of recent and imposing architecture are in this Lafayette Park district. On the edge was the monster and model power house of the Scullin street car system. Surely this was a district in which the storm cloud might make its work costly, and it did. The ruin wrought in the park was only typical of that around it. Shattered and torn by the whole day's shelling from a hundred guns, this district could not present a worse spectacle. The costly power house was a heap of ruins, with no salvage except in the form of some of the heaviest parts of the machinery. The churches were unroofed, their walls partly demolished. As for the mansions, from $8,000 to $20,000 structures, the idea of reckless cannonading can only give a conception of the infinite variety of damages.

The storm cloud passed out of the Lafayette Park district. Its north line reached Chouteau avenue, while the southern line of its path drew in toward Geyer. Before the cloud was a district of small manufacturing concerns, of many flats and of tenements, with the old rambling City Hospital, a patchwork by many builders, in the midst. Portions of the hospital stand shattered and tottering. Other portions were down flat. The 450 patients had to be cared for in temporary quarters. Perhaps more people live within this district than in any other of like size in the city. Eastward from Second Carondelet on the west, with Papin street north and Geyer avenue and Soulard street south, the population is dense. That the dead was not 1,000 and the wounded 10,000 is the mystery after one has threaded this part of the storm cloud's path. "Where is the worst?" one was asked. That was hard to tell, but certainly there was nothing worse than what could be witnessed from Lafayette Park northeast to Broadway, a long mile. It was desolation, ruin to those who had little. There were streets which were not traversed because of the debris. There were a thousand roofs gone, a hundred thousand windows smashed, whole rows flattened to the ground, street cars overturned and wrecked. Some people moved from this district at once. Most of the tenants stayed to pass the night in practical "camping out." The problem of house hunting was too much for them the first day after the tornado.

Up and down Broadway, from Chouteau avenue for a quarter of a mile each way, the storm wrecked the stores. Here a new business spirit had taken root within the past two or three years. The retail merchants of South Broadway were full of hope for that section of the city. The tornado struck them hard, very hard. It tore off roofs and it ruined stocks of goods.

IN OLD ST. LOUIS PROPER.

From Broadway to the river, through the oldest part of St. Louis, and thence up the river to the Eads Bridge, is a chapter of itself. It is a long record of wrecks of buildings that were entitled to historical tablets, buildings in which business was done by men who have been dead so long their names have passed into commercial oblivion. Crowds surged up and down the Levee all day. There was a fascination in the masonry leveled from the great bridge. There was greater interest in the big river boats driven across the river and left careened on the Illinois bank. And then, too, from this side there was a vista of unroofed buildings and jagged walls the whole length of Bloody Island.

Two facts were made most impressive by a trip over the whole path of the tornado. One was the widespread havoc wrought with the wires—telegraphic, telephone and trolley. The other was the deluge of dirty water which the storm cloud brought with it. Other parts of the city came out bright and fresh after the hard rain. In the path of the tornado proper every building shows the spattered, dingy effect of this muddy water, which must have been sucked up somewhere and then turned loose by the whirling cloud.

THE CITY HOSPITAL WRECKED.

The St. Louis city hospital was over crowded when the storm struck and demolished it and it was first thought that at least 200 of the patients under treatment were killed.

The damage to the hospital was appalling. Division M was completely demolished. The west end of division No. 9 was blown out. Division No. 10 was completely destroyed. The southeast corner of the west wing was torn out. The new erysipelas ward, under construction, was razed to the ground. The dead house and the crematory were completely destroyed. Division 6 was wrecked. The third story of the east wing at the southeast corner of the building was blown in. The amphitheater was badly damaged as was also the surgical ward.

In the darkness which followed the storm the hospital presented a desolate appearance. The rain dripped through the wrecked portions and every few moments a crash of falling plastering caused the inmates of the institution to shudder with fear and tremble with fright. The dining room had been destroyed and no one received much to eat. A hasty lunch was prepared, everyone being glad to get a bite of cold food.

To all appearances the hospital lay in the very path of the storm center, and the havoc it created at that point is beyond conception. The entire structure was unroofed; the south wall was blown out and when the acute surgical ward followed, the thirteen occupants were buried under a mass of ruins. Only one, James Dunn, was killed out-

ST. PAUL'S GERMAN EVANGELICAL CHURCH, SOUTH 9TH STREET.
(NOTHING BUT THE PULPIT LEFT STANDING.)

KOERNER'S GARDEN, 18TH STREET, NEAR LAFAYETTE AVE.

right. Of the other patients, over 100 were more or less severely wounded, and were dug out as soon as possible by Dr. Sutter and a corps of physicians.

The top story of the west wing, the whole south side of the west wing, the top story of the south wing and a big portion of the southeast corner of the east wing were blown entirely out. Besides this the stables, ice house, dead house and other adjuncts to the hospital were entirely demolished.

All the windows in the entire structure were blown out, in many cases the glass and debris falling on the patients, bruising and cutting them to a considerable extent. The rain poured in from the top and sides of the ruined structure, and the fright and consternation of the inmates of the institution can not be overdrawn, the employes state.

While making the rounds about 9 o'clock, Dr. Sutter found that a patient, Francisco Rodriguez, a servant formerly in the employ of the Mexican Consul here, had died apparently from fright. She had come to the institution several days ago to be treated for a minor complaint, and Dr. Sutter is confident that fright alone was responsible for her death. The expression on the dead woman's face, as Dr. Sutter removed the sheet which covered it, was something awful to look upon.

REMOVING THE SUFFERERS.

As quickly as possible the patients were removed from the upper stories of the hospital and deposited on cots in the two lower stories. Even these floors were uncomfortable to an extreme, in many places the water soaking through the upper floors and falling below in streams. At one time fears were entertained that the top part of the building might become watersoaked and cave in upon the patients below. Dr. Sutter said he was positive that the place was unsafe, and he issued orders that the inmates be removed from it as soon as possible. The worst of the injured were removed at once to the House of the Good Shepherd, at Seventeenth and Chestnut streets, and to the Alexian Brothers' Hospital, both of which institutions threw open their doors when they received the news of the hospital's catastrophe.

It was said at the hospital that the storm struck there about supper time. Before any one had time to realize what had happened almost the entire structure was in ruins. Lights went out immediately, and the place for about a half hour was in total darkness, save for the flashes of lightning which occasionally lighted up the surroundings. After the worst of the storm had passed, the employes could see to carry on the rescue work, and when night settled down for good the place was lighted by means of candles.

At the time of the storm the institution contained 416 patients. While the elements were raging at the worst many of the inmates became ungovernable, and broke away from their attendants and sought refuge in neighboring houses. It was impossible in the confusion to tell how many had left the institution in this way. The prisoners confined at the hospital raised all kinds of commotion during the storm, and one of them, James E. Wood, escaped. The quarters in which they were confined were not damaged, but steps were at once taken to have them removed to some point of safety.

Dr. Sutter considered it a miracle that the inmates at the hospital escaped as luckily as they did. One can hardly look at what remained of the building and think that but two persons were killed outright.

PHYSICIANS' HEROISM.

Dr. Sutter and a corps of physicians, nurses and volunteers from the neighborhood made things as comfortable as possible for the patients injured by the falling debris. The kitchens, dining rooms, offices and Dr. Sutter's private quarters were all turned into sick wards. When the physicians had finished dressing the wounds of the injured there were dozens of patients who exhibited bandaged heads, lacerated breasts, arms and faces. In several instances it became necessary to crowd two patients in a narrow bed. Invalid men and women, further weakened and terrified by the storm and their injuries, called aloud for medicine to end their misery, while others cried for something to put them to sleep. One man whose elbow was fractured leaped from his bed in pain and started to run through the hall. He was finally quieted. With the assistance of neighbors no one of the 416 hospital patients was neglected. At 9:00 o'clock at night the water had commenced to drip into the basement and second floor of the east wing, where temporary wards had been made. Dr. Lutz offered accommodation at the Alexian Brothers' Hospital for 150 patients. The offer was accepted by Dr.

Sutter, who immediately made an effort to remove all the patients from the rain-soaked and tottering structure.

"It is a terrible way to get rid of an old barn," remarked Dr. Sutter while resting a moment from his efforts, "but now the city will be compelled to erect a new and modern hospital. It is astonishing that hundreds were not killed. How the patients managed to scramble to safety is a miracle to me."

George Wilson, a patient confined on the second floor at the City Hospital, was picked up in his bed by the wind and blown into the yard below. He landed on his feet and scrambled into the basement. After he had recovered from his fright, he said :

"I didn't know what had broke loose. I knew something awful was happening, and thought may be the end of the world had come. There was a roar. The walls in our division fell out. The rain rushed in on us. Then I felt myself picked up in bed and borne out on the ground. I somehow managed to scramble to the basement, although suffering severely from rheumatism."

The acute surgical ward, which was leveled with the ground, was completed two years ago at a cost of $12,000 to the city. It was new and elegantly equipped.

The new erysipelas ward was almost completed. It was to have been a two-story brick structure at a cost of $5000. It was almost totally wrecked.

Investigation the following day disclosed still further damage. The machinery in the engine house was barely touched, but a steam pipe broke soon after the storm began and caused much excitement. The roof of the

ANCHOR HALL.

RÉSIDENCE OF DR. STARKLOFF.

boiler house caved in and caused a little damage. The carpenter shop was entirely unroofed. The old records were stored in the second story, and they are almost all lost. The few volumes that have been saved are water soaked. When the wall was blown out of Division 9, on the third floor, a man was carried with it, and landed over in Carroll street, over 500 feet from the hospital. He picked himself up and walked back to the building as though nothing had happened. The new building for the treatment of erysipelas, which was in the course of construction, was partly blown down.

The number of deaths at the hospital from the cyclone increased to four within twenty-four hours. Besides the man Dunn, who died from injuries received during the tornado, there were three deaths during the night. One was Jean d'Archimbault, who crazed with fear when the partition wall of his division fell near him, jumped up and proceeded 30 feet on one sound leg and a freshly amputated limb. He was tenderly carried back to his cot when discovered, but four hours later he expired from the shock.

Francisca Rodriguez, an aged ex-servant of Mexican Consul Sardaneta, also died of fear.

Another victim was a woman named Collins, aged 45, suffering from inflammatory rheumatism, whose death was brought about by frenzied terror at the storm.

The scene at the hospital during the tornado, with roof, walls, partitions and floors collapsing on all sides, was indescribably awful. Physicians and nurses, however, behaved with calm courage and heroism, but it was long before they could control the army of stampeded patients, many with limbs in plaster or bandages, who rushed wildly about in search of safety.

It was a miracle, considering the damage it underwent, that a fearful loss of life was averted. In many cases the divisions or wards were wiped out, patients' cots and furniture being hurled into the grounds. One man, a consumptive, was propelled from the fourth story of the west wing to the fence on the west side, fully 150 feet. He was not injured, but at once ran back into the hospital.

Desolate, surrounded on every side by evidences of the tornado's mad fury and itself a dismantled, uninhabited ruin, stood the wrecked and deserted Hospital. Its fences, its shade trees, its statuary, had been raised to the earth. Its grass plats were littered with broken slate, brickbats, cornices, telegraph poles and roofing material. Its outhouses, almost without exception, were demolished. Its wagon shed, stables, ice house, mortuary, crematory, boiler-room and laundry were no more. Its vast western wing was a pitiful wreck, roofless and piled up on every floor, with plastering, rubbish, broken cots, torn blankets and furniture. Its partitions had fallen in, its chimneys were down and in almost every division great pools of stagnant water gave evidence of the visitation of the fluid as well as of the tornadic element. A more heartrending scene cannot be pictured in any imagination than this sombre pile which two days before contained nearly half a thousand human beings and scores of public officials.

HORRORS AT THE POOR HOUSE.

The destruction at the City Poor House was also enormous, and the fright of the aged poor was something terrible to witness. With the crash of the walls as they fell like match boarding before the wind were mingled the wild shrieks of unfortunate patients, whose mental deficiencies were aggravated by the tumult of the hurricane. The situation was sufficiently horrible to unman the hardiest. The entire roof of the female pauper building was torn off and flung hundreds of yards away. The front porch of the center building was utterly demolished, the large posts being twisted away from the walls. The main damage done at the big institution, however, was caused by the destruction of the tower which stood upon the top of the central building. The tower was blown down and crashed entirely through the building down into the basement. The weight of the tower and portion of the roof was heavy enough to carry with it each of the three floors of the building situated beneath where the tower stood.

At the moment the storm struck the tower and destroyed it Mrs. Ottersen, one of the attendants, was in the chapel, which is situated in the basement directly beneath the tower. Mrs. Childress, another attendant at the institution, was in the act of descending the stairs leading from the third story, which were also directly beneath the tower. When the immense structure rocked in the wind and the tower was sent crashing through the roof, it caught Mrs. Childress and pinioned her to the staircase. She was hit by a flying brick and knocked down, which fact saved her life, as had she been able to take another step she would have been carried downward with the debris. When found she was lying on the staircase with the lower part of her body hanging over the newly created chasm. She was conscious, but suffering agony from the frightful gash in her head and the heavy timbers and beams which confined her body and crushed her chest. She was extricated with much difficulty and immediately given all the attention and care possible. At a late hour last night her condition was very serious and it is expected she will die.

Mrs. Ottersen, who was in the chapel at the time, was completely buried beneath the mountain of bricks, masonry and timbers, which crashed from above. She presented a piteous sight when the army of workmen finally located her and started to remove the debris from her body. It was a task of some time and difficulty, as several immense beams had become lodged directly upon her and had become fixed against the opposite wall. Her legs were pinioned to the ground while her chest was crushed flat by a great rafter which almost prevented her breathing. Strange to say she also was conscious when found. Her condition is more dangerous than that of Mrs. Childress, and the physicians do not hold out much hope for her recovery. Although no bones were broken, so far as could be ascertained, she has been injured internally and her body and face were horribly bruised.

There were eight of the inmates of the Poorhouse who received numerous wounds from flying glass and brick, but it is not thought that their injuries are in any way

RESIDENCE OF DR. HAUCK, JEFFERSON AVE., AND ALBION PLACE.

SCENE ON MISSISSIPPI AVE.

dangerous. Several of the attendants were cut slightly by flying splinters of glass.

The deadhouse, which had two dead bodies in it, one of which had been sent there only a half hour previous, was completely demolished by the furious wind. The building, which was a weak one, was entirely obliterated, and the two dead bodies swept away in the wind. The keeper of the deadhouse and potters field, who sleeps in the deadhouse, had left the building just before the storm and had gone up to the main building. He owes his life to the fact, as he undoubtedly would have been killed had he remained in the shanty.

PROVIDENTIAL ESCAPES.

Incredible as it sounds, with 1,300 patients in the building, 750 of whom are insane, not a life was lost at the Poorhouse during the storm. The insane patients in their ward gave the attendents a great deal of trouble during the first spell of the fierce wind. They could not be controlled and filled the building with their shrieks and cries. After the storm had abated a little, they were induced to quiet down. The new building for the male insane patients, which is the most westerly of all the buildings out there, was not damaged in the least by the storm. The engine-room suffered from the loss of the stacks, which were blown away. Hundreds of windows were torn from their casements and flung to the ground, and shutters were flung in every direction.

A view of the damaged central building presented a desolate spectacle, the diffent floors being open to the sky and dripping with floods of water. The top floor was used for the attendants, the next floor under the tower was used by Superintendent Overbeck, and the second floor by Dr. Meredith. In the basement was the chapel. Everything in the rooms mentioned is a complete loss. The damage to the buildings is estimated at at least $15,000.

In conversation with eyewitnesses of the storm at the Poorhouse it was stated that the first touch of the disastrous wind was felt at 4:50 o'clock. The patients have their supper at 4:30 o'clock and get to their floors at 4:50. It was just as the patients reached their floors that the storm struck the building. The wind came first from the southeast, then veering rapidly to the southwest it came hustling back again. The tower was torn off with a loud report and as it crashed through the roof the whole building shook to the foundations. A panic ensued in which the shrieks of the women could be heard above the wind. The maniacs added to the terror of the patients, and for a time the utmost confusion reigned. However, when it was seen that the storm was abating, order was quickly restored and the work of searching for the injured begun.

The Female Hospital escaped with little damage, the roof of the west wing being torn off, with numerous smaller damages to the interior of the building. A number of the inmates were cut by flying missiles, but there was not a single serious accident reported.

The cottage at the Insane Asylum was damaged considerably by the wind but no one was hurt.

UNION DEPOT POWER HOUSE.

The Union Depot system of electric roads is one of the largest in the world. Its tracks exceed 100 miles in length, and to secure power for its Southern divisions a mammoth structure was built some four or five years ago. It covered an entire block on Geyer, Missouri, and Jefferson Avenues. It was not a temporary structure, but was massive in the extreme, its masonry and brick work giving it a fortress appearance. Its machinery was the finest money could buy and its smoke stack reared its head so high that it was a local land mark. It was one of the highest and strongest smoke stacks ever built, and yet it fell before the winds fury just as hopelessly and helplessly as a tree with a rotten stem. A cyclist who was riding home from his work sought refuge in the power house from the rain. In common with everyone else he thought it was safe against any storm.

But he was wrong. The section of the building into which he and three others rushed fell almost as soon as they entered it. The cyclist crawled out of the wreck, bruised and battered, leaving behind the bodies of his three companions, all of whom were crushed to death. Mr. C. P. Gregory, an official of the company saw the building fall, and gave the following vivid description of the scene the evening of the storm.

"I was standing near the northwest corner of the block, in a vacant lot, and was but a few feet from the entrance to the machine shop. I saw two threatening cloud strata approaching, one from the southeast and the other from the southwest—dark and forbidding banks, resting, as it were, upon the green background. These clouds seemed to come together with an awful clap of thunder. I at once rushed toward the machine shop, yelling to the men there, about sixty in number, to get out. Just then the mighty force struck the car sheds, which are south of and adjoining the power house. The sheds were crushed in like an eggshell. There were several teams and coal wagons in the shed at the time, and two of the teams, with four horses each, were in the ruins. All the horses were killed or maimed. I don't know whether the teamsters made their escape or not, but fear that all did not.

"As the car shed walls and roof went in I yelled to the engineer, Zimmerman, who was at his post in the power house, to run, but before the poor fellow could make his escape, the big brick chimney which stands about the middle of the plant, between the two boiler rooms, came down with a crash, burying the whole power house plant beneath the debris. Zimmerman was instantly killed; we only a little while ago succeeded in getting out his body. You can get some idea of the terrible force of the elements when I say that the car shed was 400x300 feet, the brick chimney was 162 feet high by thirty-eight feet in diameter at the base, and yet the whole was a huge pile of bricks and debris in less than a minute's time. The men in the machine shop barely escaped before the car shed walls went in.

"The disaster ties up the whole Scullin system except the Bellefontaine and Mound City line, which is supplied by another small power house, which escaped injury. The first work we shall do is to search the ruins for the

GENERAL VIEW OF LAFAYETTE PARK.

RESIDENCE ON PARK AVE.

rest of the dead, if any there be. I am down-town to-night laying in a supply of tools, and we shall have a force of men at work before daylight."

A MARVEL OF RE-CONSTRUCTION.

It required ten days of unremitting labor to effect temporary repairs to this mammoth structure—not regulation working days of eight hours each—not even days of twelve hours each—but days which extended from midnight to midnight. That the damage of the storm could be repaired in even a temporary manner in so short of time may well be regarded as one of the marvelous feats of science and activity.

It is doubtful, indeed, if any corporation of the kind ever accomplished such almost marvelous results in so short a space of time. Those who saw the power-house on the morning after the storm, with its great brick smoke stack in ruins, its massive brick walls crumbled in over its machinery, its roof torn asunder and many of its cars smashed to pieces, will never forget the spectacle. Experts, even giving the company the benefit of the doubt and counting on prodigious efforts, in no case admitted the possibility of a resumption of the traffic of the roads in less than two weeks. Vice-President Harry Scullin said little, but if ever man was equal to the occasion he so proved himself. Four days after the storm a portion of the vast machinery had been cleared off and set in motion. The Tower Grove and California avenue divisions commenced running at once. A temporary roof was built over the working plant.

The huge dismantled brick smoke stack was then tackled. A force of bricklayers worked from dawn till dusk, eating and sleeping on the premises. By night another force of men labored, erecting scaffoldings so as to be ready for the bricklayers in the morning. And lo! the smokestack within a week raised its head higher than anything in the neighborhood. It was completed before the National Republicans convened three weeks after the cyclone. This stack, 150 feet high, originally took 10 weeks in building.

Sixty additional cars went into service June 7, and a horse-power of 3,200 propelled the machinery. The full power under normal conditions before the wreck was 8,000 to 10,000. The bulk of the machinery was still exposed to the elements when restarted, though cleared from all debris, and the 3,200 horse-power to be used was only secured by running the boilers to their highest capacity. They being of the very best tubular make, there was no risk in this.

By the middle of June a permanent roof had been placed over the entire power house.

On the clearing up and restarting of the machinery a force averaging 500 men a day was at work for days. Mr. Harry Scullin said on the evening of June 6:

"Give all the credit to the men. Nothing can be too good for them. They have worked with a loyalty and will I never saw equaled. Day and night they have stuck to the task that at one time seemed almost hopeless. I have never had to tell them to stay late or come early. They have been here all the time. I am certainly proud of what has been accomplished and equally proud of the spirit of loyalty the work has called forth in our men."

STREET RAILWAY SERVICE PARALYZED.

The power house of the People's Cable railway was also badly wrecked. Hundreds of cars were injured and several destroyed. The machinery escaped, but the entire second floor was wrecked, and when operations were resumed, the immense wheels revolved in mid air. The storm had moreover the effect of stopping every electric street car in St. Louis. The terrific force of the wind tore the wires of the lines from their fastenings to the side posts and in many instances blew the posts down. Every car run on the electric lines in the city was stopped almost instantaneously when the storm struck the city. The cable cars did not suffer so much. They however, were, of course, impeded by the stoppage of the cars crossing their lines.

The Olive street line was stopped by a perfect network of wires at Nineteenth street, which stretched across the entire open space at the corners. Owing to the prompt action of an employe in running to the City Hall for assistance the wires were cut away and the Olive street line resumed operation in about 20 minutes. The Fourth street cable was stopped by an immense number of telegraph poles which were blown down on Choteau avenue from Fourth street west. The poles were piled across the car tracks in all directions. The cable was shut off until they could be cleared away.

The Broadway cable ran successfully through the storm

for about five or ten minutes when the destruction of buildings on South Broadway effectually barred the street. The cable continued to run along all right, but the cars were stopped by the vast amount of debris and timbers thrown along the tracks.

The entire Lindell system was stopped by the storm. The wires were twisted and torn from the poles and every car was brought to a standstill. This was the case with every electric line in the city, save the Bellefontaine line, which was stopped for an hour and a half by the cross lines of street railways being blockaded. At the end of that time the Bellefontaine managed to resume operations.

The Chouteau avenue sheds of the Lindell Railway, were totally demolished by the storm and an entirely new structure will have to be erected.

HOURS OF HORRIBLE SUSPENSE.

The wholesale stoppage of the car lines crowded the streets with thousands of people on their way home. Every doorway and entrance was jammed with frightened men and women in momentary danger of being killed by the hundreds of live wires, which hung about in all directions. The Olive street line, when it resumed operations after its 20-minute stop, was so crowded with humanity that movement within the cars was an impossibility. People got on the cars as far west as Twelfth street in order to obtain standing room even when the car returned on its west-bound trip. Ladies, running risk of being thrown to the ground and killed, frantically caught on to the cars while they were in motion in their endeavors to escape

EXCELSIOR LAUNDRY.

ST. LOUIS JAIL.

from the horrors of electrocution, which threatened everyone on the streets. In many instances ladies and young women were thrown down and narrowly escaped injury. That no serious damage is reported from this direction is a mystery. Inquiries among the employes of the various roads down town elicited the fact that very few, if any, accidents occured to passengers.

The day following the storm saw fewer street cars running than on any day for years, the service being poorer on the whole than in the old horse car regime. Only two lines on the Scullin or Union depot system were in operation. The Benton-Bellefontaine ran right along and it was the only street car line in the city that was not disabled by the storm. The Mound City resumed operations early in the morning. All the lines running to the Fair Grounds were running except the one on Jefferson avenue and Mr. Scullin's Fair Grounds line. The cars of the latter were used on the Mound City road. The Union, Northern Central and Cass avenue lines all started up about the same time next morning, and the Franklin avenue line at 8:30 the fatal evening. The only other line operated by the Chicago syndicate is the Southwestern, which parallels the Broadway line of the same company. The Southwestern started up a day or two later and proved of immense service to a large number of people living south of Arsenel street in the territory covered by some of the Scullin lines. The Southern electric also expected to resume operations on a part of its route. It runs to Carondelet and is ordinarily patronized by thousands of people living north of the Wild Hunter's.

The Grand avenue line was operated from the water tower as far south as the Grand avenue bridge only. The service will be extended south of Mill Creek. Another line for the Southsiders which was able in a day or two to run cars was the Lindell company's new Park avenue electric road. It leaves the Compton Heights line at Nebraska avenue and continues west on Park avenue to Vandeventer, and then runs south to Tower Grove Park. The company's power house on Park and Vandeventer avenues was started up at noon Thursday, but the company's lines south of Mill Creek Valley were nearly all down and the power was used by the Page, Spalding and Delmar avenue divisions of the Lindell system, all running to the West End.

The Olive street cable did a tremendous business. The electric lines operated by the same company, on the other hand, fared badly. The power house from which they were operated was partially demolished and it was several days before traffic on them was resumed. The Jefferson avenue line was in the same fix.

TELEGRAPH WIRES DOWN.

For several hours after the hurricane St. Louis was cut off from all communication with the outside world. Every wire, north, east, south and west, was reported down.

The telegraph and telephone services and the electric railway lines were crippled, and all street railway traffic was temporarily suspended. The damage to the wires was estimated Wednesday night at over $100,000.

The hurricane snapped hundreds of telegraph poles

throughout the city like twigs, and a perfect labyrinth of the deadly overhead wires were precipitated to the ground. The casualties from this source were miraculously small, in view of the danger to which thousands of people were exposed. As it was, the twisted wires were a constant menace to pedestrians, and scores of people are known to have come in contact with live wires. No fatalities were reported from this cause, however.

This was due to the fact that all the electric car lines shut off their power as soon as they received word that the wires were down.

At 8 o'clock Kansas City was reached by way of Mexico, Mo., and later Chicago was communicated with by way of Kansas City. The wires on the Illinois side were down for miles.

Manager Bohle had gangs of men out in all parts of the city all night, but the damage was so great that the men were unable to do much before daylight. The telegraph service was crippled for some time on account of the storm.

The telephone wires, with the exception of a portion of the West End circuit, were all down, and the downtown districts were without telephone service the next day.

A FATED CORNER.

The corner of Seventh and Rutger streets may be described as the charnel-house in the path of the storm. Thirteen bodies were taken from the ruins of two buildings on this corner within twenty-four hours of the storm, and there were still forms under the heaps of brick and stone that marked the place where the buildings stood.

A boarding-house stood at the southeast corner. When the storm reached it the walls collapsed as though made of stiff dough, and the unfortunates inside were entombed in the twinkling of an eye. The high wind that swept around the corners after the main storm passed created an eddy over the ruins into which was tossed a mass of debris from the entire neighborhood. The work of rescue was difficult and necessarily incomplete. People in the neighborhood say there were 40 souls in the house when it went down, and but nine corpses have been taken out.

On the opposite corner from the boarding-house searchers for bodies found four mangled trunks and sent them to the morgue. A man half daft with terror assured Sergeant Meehan that the body of his wife is buried there. He was to excited to give his name, but neighbors said he lived in the house, and that the woman has not been seen since the storm.

One might travel miles and find few more desolate, heart-sickening pictures than that around Seventh and Rutger streets. A pile of brick and broken timbers stretched from curb to curb on the Rutger street side, east of Seventh. Along the sidewalk on the north side of the street a pathway was being made through the debris, and through this the few who had business within the ropes that guarded the graves of families were allowed to pass by the police. Upon a mass of tangled wreckage, in plain view of the curious mob, an old woman sat all day surrounded by a few of her battered household goods, dug from the wreck. About her were gathered a few

WRECK OF THE STEAMER "HENRY SACKMAN."

OTTENAD FURNITURE CO., BROADWAY AND SOULARD STREET.
(A NUMBER OF PEOPLE WERE KILLED HERE.)

friends trying to persuade her to go away. But she would not go away. For years and years she lived on that corner. Her children were born there, and from the door of her house she followed the body of her husband a long time ago—so long ago she cannot remember all about it.

She expected to die there in peace and quiet with her relatives gathered about her bed, but from the expression in her sad, dim eyes, as she gazed on the sickening ruins, she would have been satisfied had she gone down with the rest, and never known what caused the candle of life to go out. And so she would not go away but remained for hours, a pale, solitary figure, pathetic and pleading in a vista of woe and sorrow.

The buildings in this vicinity were totally wrecked. They cannot be repaired. They represented to the honest people who occupied them before the storm, the savings of years, and that was being swept away in the space of a minute was beyond their comprehension. They were dumb with vague surprise and went about unable to realize the full extent of their affliction. It was like a holiday. All the stores were closed and thousands lounged in the streets. Men and women stood moodily in sheltered places watching the crowds. Some of them had babies beside them and the prattle of the little ones sounded strangely out of place in the surroundings. On every hand were evidences of wreck and ruin. But a saloon within 50 feet of the place, where the nine bodies were recovered, was doing rushing business and shouts of drunken roisterers, mingled with the clink of glasses and the jingle of money, resounded mockingly in the ears of the stricken and homeless.

A COMBINATION OF HORRORS.

Wind, lightning, fire and rain formed the hellish combination that slaughtered and maimed and rendered houseless and destitute scores of defenseless residents of the district bounded by Park avenue on the south, Laclede avenue on the north, California avenue on the west and Dolman street on the east. The entire section presents a sickening picture of desolation, ruin and misery. Amid the wreck of palatial residences, handsome flats and cottages less pretentious, the dead and dying have been taken out here and there throughout the entire district, and every block has horrors upon horrors to relate.

In many instances the awful work that was begun by the tornado's crushing force was made doubly terrible by fire, and when the elements had spent their fury the flames blackened the wreckage in token of the ruin that had been wrought. In more than one case people who were pinioned beneath fallen walls and debris were burned to death before the eyes of family or friends who were powerless to aid them, and many others would have died equally horrible deaths but for the bravery of the rescuers.

Nothing that lay directly in the path of the storm seemed capable of withstanding the terrific force of the wind. Great stones weighing thousands of pounds were carried for blocks and dropped wherever the wind seemed disposed to leave them. Some of the most substantial buildings in town were leveled to the ground, while perhaps an old dilapidated structure next door escaped without injury.

Probably the most remarkable case in point is the destruction of the power house of the People's Railroad, (Fourth street cable), at 1801 Park avenue, already referred to. It was a massive structure, only three stories in its tallest part, and built of pressed brick and stone. The walls ranged from 20 to 30 inches in thickness, yet they were crushed as though they were built of paper. The entire top story of the main office building was thrown into the street, and the engine room was unroofed and wrecked. The damage was $100,000, to say nothing of the loss that accrued from the suspension of traffic on the road.

When the roof was lifted off great stones and a perfect storm of massive timbers were rained down into the engine room. The big steam pipe leading to the engine on the west side was broken and every car on the road stopped almost instantly. The roar of escaping steam added to the confusion and for a time it was thought everybody in the place would be killed. The engineer, D. G. McCarron, was in the engine room and had a narrow escape from death. The car sheds on the block to the south were unroofed and partially demolished, and a number of cars smashed into kindling wood. Manager Mahoney put a force of 100 men, mostly employes of the road, to work cleaning away the debris and getting in shape to rebuild the power house.

HANDSOME RESIDENCES DESTROYED.

From there all the way out Park avenue to beyond Nebraska, almost every other house is more or less damaged, and a large majority of them are total wrecks. This is particularly true of the magnificent residences to the north of Lafayette Park.

The Soderer house, on the corner of Missouri and Park avenues, was badly damaged. The stable was a mass of ruins, and a fine horse and its driver, a colored man named Taylor, were buried under the debris. Both were killed. Julius Houck, brother of the Doctor, who lives at 2320 Park avenue, has a house that is not fit to live in. Two flats adjoining, owned by his mother-in-law, are a total wreck.

The house on Park avenue, near Twenty-second street, where Alexander Selkirk boarded with his wife, is a mass of ruins. Mrs. Selkirk lost everything—handsome ornaments, jewelry and her clothes.

The house of Jerome Hill, the cotton factor, was dismantled, with great gaps everywhere, the top gone, and lightless windows. All the family were at home the next morning removing furniture, and trying to save from the debris what was savable.

John Endres, who lives next door to Mr. Hill, directed the moving of his belongings out of his devasted place. In spite of his loss, Mr. Endres good-naturedly said, he was only glad nobody in his house was hurt. He had spread a luncheon on the parlor mantle, a couple of bottles of wine from the cellar had been saved, and he insisted upon everybody in his household taking a bite to

SEVENTH AND RUTGER STREETS.
(TWENTY-EIGHT PERSONS WERE KILLED HERE.)

RUSSELL AVENUE, EAST OF OREGON AVENUE.

sustain their courage. His four daughters and the two Misses Wamsganz, sisters of Mrs. Endres, went to stop with their grandmother until Mr. Endres could make other provisions for his family. He slept part of the night on the floor of the Carr residence, next to his own, and, like a faithful watchman, got up every half hour to see if his chattles were all right. Mr. Endres' courage and philosophy over his loss, did much to inspire his neighbors with the same sort of feelings.

Mrs. James Carr, relict of the late Judge Carr, with her sons and daughters, lives next door to Mr. Endres. Her house was the least damaged of any of those on that side of the street, though there was not a whole pane of glass left in it. Mr. James Carr, Jr., was busy putting in the broken windows, after having spent part of the night in helping Mr. Endres to bring a little order out of the chaos. Mrs. Carr housed a number of bleeding and fainting neighbors, and beds were made all over the floor to take care of the homeless. She attributes the saving of her house to the solidity with which it is built, but equally strong houses proved the fallacy of such reliance.

The Misses Carr had just returned from the wedding of Miss Fisher to Mr. Leon Fasset in the little Church of the Unity in Benton place. The wedding guests had barely left the church when the roof blew off, and they all ran for their lives. Miss Carr has in her possession two mackintoshes and four umbrellas which she found in the church, and has taken care of, awaiting their owners. The Carr family sought refuge in the cellar between two large brick piers, which swayed and rocked, but did not give way. The wind tried itself in the Carr residence without doing great damage, but it tore frames from doors, broke locks in half, twisted one hinge off a door and left the other on, shattered some window panes, while it scratched others in zig-zag fashion, drove enormous boards through one side of the house in the third story, and never moved a wooden bucket from an unsheltered position in the yard.

SAW HER CHILDREN BURN.

Horror's crown of horrors was reached at the home of John Bene, 1400 Ohio avenue, just off Park avenue, where the house was first wrecked and then burned. Mrs. Bene and her two children, Sylvester, aged 5, and Oliver, her two-year-old babe, were buried under the debris, and the latter was burned to death before his mother's eyes. Sylvester, the next older child, was also badly burned, but a corps of rescuers succeeded in digging down to where he lay before life was extinct. The poor little fellow was unconscious and his skull is fractured in addition to the severe burns he received. Mrs. Bene could see her little ones from where she lay pinioned under some heavy timbers. The wreckage was on fire all around her, and she was being slowly roasted to death, but seemingly oblivious to her own danger and the excruciating pain she must have been suffering, she implored the rescuers to cease their efforts to reach her and save her babies instead. The maternal instinct in her was so strong as to overcome all physical pain. When she was taken out she was found to be badly burned about the legs, hips and feet, and bruised from head to foot by falling bricks and timbers. When she learned that one of her babies was dead and the other probably fatally injured she fainted.

Five people were instantly killed by the demolition of the home of the Schwerdtman's, at Ohio and St. Vincent avenues. The house was wrecked till not one brick remained upon another, and the entire family was buried beneath the ruins. C. Schwerdtman and his old mother were found side by side, dead. The bodies of Mrs. H. Schwerdtman and her two children were taken from under the debris an hour later. Mr. Schwerdtman, husband of the latter, was on his way home when the storm broke. When he reached there and found the house in ruins, and his wife, children, mother and brother, dead, he was almost frantic with grief, and his friends feared for a time that he would lose his reason.

Two blocks further north, at 2700 Caroline street, the grocery store owned by Anton Rubald, was buried under the wreck of the three-story building in which it was located. Rubald sought safety in the cellar, but his head and back were severely cut and bruised. The grocery stock was almost a total loss. Mrs. E. Filsinger's home, at 2635 Caroline, was blown down and she was badly hurt. She was taken to the Rebekah Hospital.

HOMELESS AND DESTITUTE.

When the first grey streaks of dawn began to break on the eastern horizon, the terrified residents of the district at the south approach of the Jefferson avenue bridge resumed their search for the dead in the wrecked tenement at 814 South Jefferson avenue. The corpses of Mrs. Claypool and her daughter Cora, were taken out during the night. Two and possibly three more bodies they knew were buried under the mass of brick and timber in which they had delved till after midnight. By 9 o'clock the remains of old Mrs. Cahill were reached and a little later Ethel Claypool's crushed and bloody body was taken out. They were sent to the morgue. In the same house Baby Barnett, aged 3, had her back broken, and her aunt, Mrs. Holland, was seriously hurt. A man named Irwin, living at 2712 Chouteau avenue, had his head crushed in a carpet factory at Twenty-second and Chouteau.

LaSalle street, a short thoroughfare running from Eighteenth to Jefferson avenue, is probably the most Godforsaken looking place in town. The storm injured a number of people living there and left nearly every family along the street homeless and destitute. Their misery was increased by the inability of moving vans to cart the remnants of their household effects elsewhere on account of hanging wires. Annie and Lydia, the two little children of John Hubli, living at 2313 LaSalle, were both injured by the roof falling in on them. Hubli's belongings were blown into the street, as were those of Mrs. Prack, a widow living up stairs. A bureau was blown out of one of Mrs. Prack's windows. George Weinter's furniture was scattered all over the street and broken to pieces. Nearly every family living in the neighborhood suffered a similar loss and many of them will be entirely dependent upon charity for some time.

MOUNT CALVARY EPISCOPAL CHURCH.

WAVERLY PLACE.

THE GREAT CYCLONE. 139

The most serious injury reported was that to Mrs. Ackert, 2313 LaSalle, whose back was broken by flying timbers.

Wallace Buckner and Guy Harvey, both colored, living at Jefferson and Chouteau avenues, were killed by the collapse of a brick tenement. Mrs. Abbie Woods, colored, was seriously injured by falling walls at her home, 2321 Papin street. Gratiot street from Eighteenth street east to Sixteenth, is lined on both sides with warehouses, all of which were razed to the ground. The block between Chouteau avenue and Hickory street on Armstrong avenue is probably the most complete wreck in the city. Every house, some of them handsome two-story affairs, was blown down and the street rendered impassable by bricks, timbers and broken roofs. Not a house escaped.

Selkirk's storage house at Eighteenth street and Chouteau avenue was badly wrecked. The roof was taken off and the rear wall torn out. R. U. Leonori's warehouse, at Jefferson and Chouteau avenues, which was filled to the roof with furniture, is in ruins and the contents are badly damaged. The roof was blown off the St. Louis Coffin Company's plant at Fourteenth and Poplar streets, and the top story of Butler's warehouse at Eighteenth street and Chouteau avenue is gone.

HORSES KILLED IN HARNESS.

The Lindell Railway car sheds and repair house at 2330 Chouteau and extending across LaSalle street, were badly damaged and half a dozen cars were broken to pieces. An employe named Bryant, living at 2615 Rutger, had his arm broken. The car sheds of the Jefferson Avenue

Line, adjoining the Lindell repair shop, were partially wrecked and two cars were destroyed. The wires are down and the line will not be running before Sunday.

The engine house at Eighteenth and Hickory streets was razed to the foundation. Two fine grey horses were killed as they stood side by side in harness ready to respond to an alarm that had just come in. The engine was out and there were only two or three firemen in the house at the time. The St. Louis Wire Mill is in ruins and so is the Oriel Glass Company's plant, just west of it. Over a thousand panes of glass were broken there. The vacant block to the south is strewn with wreckage of every description. The National Brewery, at Eighteenth and the railroad tracks, was partially unroofed. Singularly enough the tall shot tower, across the tracks at Twenty-second street, was unscathed, while Osborn's harvester factory, near by, had its roof blown off. L'Ouverture school was so badly damaged that it had to be closed up.

DESOLATION AND DESTRUCTION.

Looking south from the Jefferson avenue bridge was an unbroken panorama of wrecked homes, business houses, broken street cars, prostrate telegraph wires and overturned sheds. The street was filled with rubbish of every description and the scene of desolation and destruction spread as far to the south as the eye could see. It would furnish columns upon columns of most frightful recital in itself, were it not for the fact that such an area was covered by the storm that what happened to a single street was dwarfed by the appalling nature of the whole.

Beautiful Benton place, with its homes and garden-like surroundings, was wrecked from center to circumference. On this place was the handsome residence of Anthony Ittner, which stood on the corner. It was totally demolished. Miss Isabel Ittner's bed blew out of the front window in the third story and was found, after a weary search, on the sidewalk, under a lot of rubbish. The next house to the Ittner mansion was that of Mrs. Valle, which was destroyed, as was the delightful home of Mrs. L. B. Baker. Mrs. Baker, with her little daughter, had just returned from down town, and had not been in the house long when the furious elements began to revolt. The whole family huddled under a kitchen table, clinging together and comforting each other as best they could. It will cost $3,000 at least to rebuild the house, to say nothing of the furnishing. The piano is a total wreck, but Mrs. Baker's handsome saddle horse was saved. The residences of Mrs. Rainwater and Mrs. Haydock fared a similar fate.

LOSSES TO BUSINESS CONCERNS.

The following business losses were noted in the small districts bounded by Eleventh, Dolman, Lafayette avenue and Market street.

The building occupied by the Enno Sander Mineral Water Company, 125 to 129 South Eleventh street, was damaged to the extent of about $1,000 and the stock about $300.

The Purina Mills, on the southwest side of Twelfth street bridge, were completely destroyed. The building, owned by the St. Louis Bagging Company, was an old structure, worth probably $5,000. The stock of purina, owned by the Robinson-Danforth Commission Company, was utterly disintegrated. The damage to it will be almost $5,000. The company will at once engage new quarters and there will be no delay in their business.

Two girls employed in the mills remained in the building throughout the tornado and, while the walls around them were rent asunder, they escaped uninjured. An aged horse, kept around the premises for the sake of old acquaintance, was injured in the back and died.

M. M. Buck's warehouse, 810 to 826 South Twelfth street, was damaged about $1,500.

SWAMPED BY THEIR NEIGHBOR.

The Shickle, Harrison & Howard Iron Company's big plant on Twelfth street, rear of Chouteau avenue, suffered injuries that will cost about $10,000 to repair. Much of this was done by the unwelcome advent of a portion of M. M. Buck's warehouse from the other side of the street. The machine shop was wrecked, the hay barn and pattern shop badly twisted around, the stable was blown down and the roof of the foundry taken off. The 500 men employed escaped uninjured despite the flying brick bats and other debris.

The Haydock building, on Fourteenth street, near Chouteau avenue, was damaged about $10,000. Part of the ornamental center of the Fourteenth street end fell to the street, while the roof and part of the rear walls crumbled in. The loss on stock will be to the Haydock

ST. LOUIS COTTON COMPRESS WAREHOUSE.

GUNDELOCH THREE-STORY TENEMENT HOUSE.
(ELEVEN PEOPLE WERE KILLED HERE.)

Brothers' Carriage Company about $500. The Bauer-Walter Carriage Company and Stromberg, Kraus & Co., trunk manufacturers, who occupy portions of the structure, escaped with slight damages.

The Griesedieck Artificial Ice Company, at Thirteenth and Papin, lost their three smokestacks, which, besides partially crippling the works, caused a loss in the vicinity of $1,000.

TREMENDOUS DAMAGE.

The superintendent of Station C, Laclede Gas Company, on South Fourteenth street, held up his hands when asked as to damages sustained by the plant. "One hundred and fifty thousand dollars will not cover it," he replied. "We are torn up in all directions. Our Pintsch plant is only slightly hurt, say about $500, but one of our storage holders is wrecked. All the gas has escaped, our buildings are, some of them, demolished and we are in a bad way." The huge circular storage holder was 40 feet high when struck. It quickly settled down almost to the level of the ground. One man, Frank Benson, was injured by flying missiles.

The St. Louis Coffin Company, who conduct their business in two immense buildings at Thirteenth and Poplar street, lost part of the upper story of one and the roof of the other. With the damage to stock etc., the loss will, it is estimated, be over $10,000.

The Lafayette Park Stable, operated by Heier & Wiebush, was, like everything else around the City Hospital, ruinously handled by the tornado. The loss on the buildings, which were practically demolished, will be about $10,000. That on the plant is still greater. It includes one hearse, six carriages and light buggies, nine surreys, 13 vehicles of all kinds belonging to boarders and a stock of coffins and hardware. Everything was split into matches. It cannot be replaced under $30,000, making a total damage of $40,000.

H. Reemt's grocery, and another building owned by him, at 1800 South Fourth street, were damaged fully $3,000.

Mrs. Walker, who owns four houses and stores on the east and north sides of the City Hospital, loses about $4,000 on them.

DAMAGE TO HOMES ON ONE STREET.

Dolman street, going south from Chouteau avenue, presented an awful appearance. Up till the fatal night it was a pretty shaded thoroughfare, with comfortable residences on either side. Next day its shade trees lay prone in all positions across the street or the front yards of the houses. Telegraph poles and wires, roofing, bricks plastering and lumber help to make the scene one of desolation. From Chouteau avenue to Lafayette scarcely a house escaped injury and almost all were badly wrecked. Among the latter were:

Nos. 1201 and 1203, owned by George Peisch, and demolished to the extent of $3,000. Both were occupied by families who deserted them for safer quarters before the tornado came.

No. 1221, completely destroyed; value $3,000;

owned by Conrad Soehlmann. This was a two-story brick, rented by Mary Reardon, who was partially buried in the ruins, but was rescued by neighbors. Her daughter is employed as cashier at Philip Roeder's stationery store, on North Fourth street. After the storm, Miss Reardon walked towards home, little dreaming of the awful sight that was to greet her eyes. Mrs. Reardon was painfully cut. Their furniture, valued at about $600, was ruined.

Nos. 1223 and 1225, also owned by Conrad Soehlmann, was damaged $1,000.

Nos. 1227 and 1229, owned by Mrs. Koch, were simply obliterated. Loss $9,000.

Nos. 1304 and 1306, owned by Mr. Howe, were damaged $2,000.

Mrs. Lange, a widow, owned three houses opposite, which cannot be repaired under $5,000.

No. 1315, a stone-front two-story house, owned by Mr. Kiesewetter, was damaged $4,000.

Mrs. Kraft owned and occupied No. 1317, which was badly torn apart. Damage $3,000.

C. E. Miller owned No. 1431, which lost the roof and windows, worth $1,000.

Klauber & Son's property at 1435, was wrecked to the extent of fully $2,000.

Mr. Nelson, the contractor, suffered a loss of about $3,000 on his place at 1436 Dolman.

Nos. 1420 and 1422, owner unknown, are demolished in part. Loss estimated at $2,000.

No. 1613, owner unknown, will cost at least $1,500 to straighten out.

Almost every badly damaged house on Dolman street was deserted at once. As fast as they could find other quarters, families moved out. Scores of furniture wagons were being loaded up there the day after the saddest of all sad days.

AROUND THE CITY HOSPITAL.

On Lafayette avenue at 1556 to 1564, a two-story block, owner not known, lost almost all its upper floors. The loss exceeded $3,000.

Nos. 1622 and 1620 South Thirteenth street were just half gone. Damage, $1,500.

At 1411 Carroll street the entire upper portion was swept away, and the occupants, who had fled down stairs, were in a mournful fix. Loss about $1,000.

The entire front of 1410 Dillon street was blown off. Damage, $1,000.

That historical landmark known as "Cracker Castle," at St. Ange and Chouteau avenues, owned by Major Charles E. Pearce, and occupied by Mr. Frederick Spies, was struck by lightning during the tornado. It was once almost the handsomest and best appointed residence in that section, but the storm has sealed its fate. Its large tower was hurled onto the western portion of the building, which it demolished. The loss would be in the neighborhood of $3,000, but the castle will doubtless be torn down.

The upper part of 1025 St. Ange avenue was borne away by the storm. Loss over $1,000.

EAST ST. LOUIS GAS WORKS.

PLOEHN'S FURNITURE FACTORY, BROADWAY AND MILLER STREET.

MANY DEAD HERE IN A WRECKED TERRITORY.

Search for the bodies of the victims of the tornado, supposed to be still on the vacant lot opposite the old market was tedious, owing to the small force of men at work. The ruin of the market was most colossal and complete, and it was owing partly to the extent of the area to be dug over that the work of recovering the bodies progressed slowly.

Sixteen men, under the direction of George Dace, worked all day and far into the night in the ruins. The next morning the body of George Howard was dug out of the debris at the extreme west end of the market. Howard was connected with the circus which had been exhibiting on the vacant lot opposite the market, which was blown out of existence.

An immense crowd of curious people watched the searchers at work and interfered greatly with the workmen during the evening. The workmen were divided into three squads, located at three different points in the ruins near the center of the market. At 9 o'clock at night several of the workmen, who had reached the basement, thought they could hear a faint cry for help from under the debris on the northern side. The news was quickly communicated to the crowd, and a hush of expectancy fell upon the spectators as the men worked like beavers in their haste to reach the point from which the cry was supposed to have come. Diligent search failed to reveal any signs of a human body, and although the searching party shouted lustily into the ruins, no answering cry responded.

NOTHING WAS SPARED.

On South Twenty-first street, near the Mill Creek Valley, desolation was universal. A photograph of the scene from 300 South Twenty-first street to the south end of the bridge would tell the story better than words. Not a single building, telegraph pole, tree or sign seemed to have escaped. The street and sidewalk was an almost solid mass of debris—bricks, broken telephone and telegraph poles, tin roofing, water spouts, broken signs and broken glass, while the wires covered the whole in an interminable netting, just high enough off the ground to catch pedestrians across the face and under the chin, a few being low enough to trip one's toes by way of variation.

At 310 South Twenty-first street, the J. F. Swift & Co. fresh meat establishment is located, and just here the hurricane began business for the first time on Twenty-first street. Not a single window in the building remained intact, the whole place being flooded.

The stock of $3,000 worth of meat was covered with water, bits of broken glass and flying mud and sand. The managers, however, made necessary deliveries two days later. The damage to the building was $500.

Officer Bart Keaney, on the Twenty-first and Clark avenue beat, deserves much credit for the heroic work he did in assisting to rescue the injured from the wrecked store at 2102 Clark avenue. When the storm struck the store it went down without warning. Bridget Gunn, 2722 Sheridan; Gertie McKenna, 2108 Eugenia street, and John O'Connor, of 2109 Adams street, were in the store at the time, and were buried under falling timbers and brick walls. Miss Gunn's legs were both broken, left arm broken between wrist and elbow and skull fractured. She was sent to the City Hospital.

Gertie McKenna was injured internally.

O'Connor has a broken arm, a scalp wound and internal injuries.

Grone's brewery, at Twenty-second and Clark avenue, was almost totally wrecked.

The Scandinavian Church, southeast corner Twenty-third and Clark, was picked up and turned completely over.

At the Gates Wire Mill, Twenty-first and Gratiot, a scene indescribable was met with. An immense portion of the roof of the building lies piled up high on the south end of the bridge, while all around a mass of wires, poles, roofing and tons of other debris made the vicinity utterly impenetrable. Mr. Lippies, who was employed in the Gates mill said that four were badly hurt, two fatally. Mr. Lippies could not explain how it happened that of the 300 or more boys and men who were in the building only four were hurt, but he explained that he thought the vivid flashes of the burning electric light wires under the Twenty-first street bridge proved to be a miraculous warning, causing many who might otherwise have been crushed to death, to flee to a place of safety.

Mr. Lippies himself ran out of the building to see what the trouble really was, and, no sooner had he entered the open yard east of the building, than he was picked up bodily and carried over the companys woodshed, landing in his own back yard.

He was dazed, but not hurt, and, jumping to his feet, entered his back door and ran into the arms of his wife before he knew what had really happened.

The Sawyer Manufacturing Company, 1819 Chouteau avenue, suffered greatly, the building being almost totally wrecked. Four people were injured.

The Eden Publishing house, at 1716 Chouteau, was also a heavy loser. The building was unroofed and the stock damaged.

Gieren's drug store, on the corner of Twelfth and Chestnut, was badly smashed.

St. John's Church, on Hickory street, lost its steeple.

The Liederkranz Hall, Chouteau avenue and Thirteenth street, was unroofed and the entire building flooded with water.

Alexander's saloon, Eighteenth and Chouteau, was considerably damaged.

The Brown Tobacco Company, across the street from Alexander's saloon, lost the top story of its building.

Spilker's drug store, 1801 Chouteau, suffered severely.

The south wall of the nail department of the St. Louis Wire Company was wrecked, fire breaking out a little after 10 o'clock.

Fire also helped to complete the destruction of the Gates Wire Mill, breaking out immediately after the storm.

SEARCHING FOR THEIR HUSBANDS.

The district bounded by Market, Chouteau, Broadway and the wharf was a complete wreck. The streets were filled with wreckage, the upper stories of the old build-

ELEVATOR AT CHOUTEAU AVENUE AND LEVEE.

EAST ST. LOUIS ELECTRIC POWER HOUSE.

ings were blown in and the roofs of many buildings carried away. Many people were injured by flying bricks, wood, signs and the like, and were treated at the neighboring drug stores. A few of them were taken to the hospital and thence to the Good Shepherd Convent. It was claimed that quite a number lay buried under the demolished buildings, but it required days to tell the tale, as nothing could be done in the storm and darkness of the night toward searching the ruins. It was impossible to get the ambulances east of Third street, although it was reported that a number of dead and injured were lying in the district east. The drivers were unable to get to them, try how they would.

The broken telegraph and telephone wires and poles impeded progress everywhere, so that no investigation of the casualties could be made. Everywhere desolation and ruin greeted the eye. Tenement houses were caved in and the unhappy tenants were shivering in the cold, chilly rain on the sidewalks. Frightened women clustered about the windows of the great manufacturing buildings, crying and praying for deliverance from the catastrophe occurring on all sides. The men dashed here and there in great excitement, as the thunder crashed and the lightning flashed, and tried to quiet the fears of the women. It was an exciting scene in this great wholesale and manufacturing district while the storm was at its height.

GAS WORKS WRECKED.

As the storm was at its height the gasholder at station C of the Laclede Gas Company, located at Fifteenth and

Gratiot streets, collapsed. There was no noise or explosion, but the neighborhood was terrified by the sight of columns of burning gas leaping high in the air. The tank was nearly full at the time, and the full force of the wind caught it and turned it over. As the immense iron columns, which supported the tank proper fell, the gas was ignited and burned for a few moments at a fearful rate. Nearly all of the columns fell in the grounds of the company. They are 115 feet long.

One of the columns fell across Fifteenth street, and pinioned a young colored girl by the legs. It was two hours before she could be extricated.

FALLEN BUILDINGS EVERYWHERE.

Proceeding south from Chestnut street on Third a scene of woeful devastation met the eye on either side. The street was strewn with fragments of roofing, chimney tops, broken glass, bricks and telegraph poles, while the wires were coiled in every direction. In some places the huge demolished poles blocked the street; wagons broken to pieces or upset were numerous. The frightened people had in many cases taken refuge in the lower stories, sometimes in the cellars of their homes, fearing a fresh outbreak, as the lightning continued to flash, the wind roared and the water still poured down in drenching torrents. Scarcely a house escaped injury. Scarcely a pane of glass remained intact. Water flooded almost every domicile, and men, women and children, who retained their practical senses, busied themselves endeavoring to secure dry quarters for the night.

THE GREAT CYCLONE. 159

The factory of the Dodson-Hils Pickle and Sauce Manufacturing Company, at 732 South Second street, was very badly wrecked. Corner saloons and stores seemed to have come in for the worst effects of the destructive element.

A glance at the big elevator of the St. Louis United Elevator Company, near the river, at Chouteau avenue, revealed the fact that the entire upper portion had been blown away.

The mammoth establishment of the N. K. Fairbank Company, soap manufacturers, at Third and Convent streets, suffered severely. Its tanks and outbuildings were scattered in all directions. Its soap was demolished, its windows blown out and its stacks of barrels dispersed over the adjoining country.

UNION CLUB WRECKED.

The Union Club's home is, or was, one of the handsome pressed brick buildings which go to make the Lafayette park vicinage architecturally beautiful. This magnificent building was completed about two years ago. Its apartments were spacious. It had handsome parlors and smoking rooms; billiard and card rooms; in fact, all the accommodations for a modern clubhouse. The building was three stories high and about 80x120 feet, and its lower walls are now standing and the loss is almost total. Within 100 feet of the clubhouse is the Lafayette Park church and across the street from it towards the west, is the Protestant Episcopal church, also a wreck. In none of these public buildings did a death result from the storm, although Mr. Michel, the steward at the club, received severe bruises.

Of all the homes on Lafayette avenue, the Duestrow homestead has in recent months attracted most attention. Like other homes in this neighborhood, it is a wreck. The building is four stories high, built of stone, and one of the handsomest residences in that part of the city.

Next to the Duestrow homestead is the residence of Mr. G. E. Wetzel, at 2327 Lafayette avenue. The Wetzel home was also an attractive residence and is now almost a total wreck. It is occupied by Mr. and Mrs. Wetzel and Mr. and Mrs. Ringer.

Before telling more about the marvelous escapes in this section of the city, an observation which will almost universally apply should be made. That the loss of life in the dozen blocks which surround Lafayette park did not run into the thousands is due to the fact that the people rushed to the basements. Hundreds of families saved their lives by this means. They seemed to know intuitively where the safest room was located and sought it at once where there was no basement, without being able themselves to tell just why they did so.

LIVES LOST IN A FIRE.

Fear was added to the fury of the storm by an alarm of fire from the great plant of the St. Louis Refrigerator and Wooden Gutter Co., 1200 South Main street. It was a four-story building, magnificently equipped. The hungry flames ate up the interior and gnawed at the walls with impunity, for there was only one fire engine in

DOUGLAS SCHOOL, EAST ST. LOUIS.

RESIDENCE OF DR. E. PREETORIUS, PARK AND MISSISSIPPI AVENUES.

sight, and it looked like a pigmy battling with a giant. The network of wires blown down was so complete that the streets were impassable, and the Fire Department became powerless.

The flames fanned by the fury of the atmospheric eruptions, cavorted like demons and wrapped their liquid tongues around the wavering walls. At last they fell with a crash that was heard and felt for miles and the fire fiend was satisfied.

The material damage was estimated at $200,000.

The Aluminum bicycle factory at Main street and Park avenue, belonging to the same company, was unroofed but escaped the flames. Many lives were lost in the building.

The factory employed over 300 hands, and as they were still at work when the crash came, followed by the fire, it is marvelous that any of them escaped.

A HUGE FACTORY DESTROYED.

One of the greatest freaks of the wind was the havoc wrought at Liggett & Myers tobacco factory in the west end of the city. This plant is to be the largest in the world of its class, and consists of some thirteen buildings grouped close together. All lie east and west and apparently directly in the storm's progress. The natural inference would be that all were damaged to an approximately equal extent, but the effect was far different. The buildings in the south group were practically completed, but escaped entirely, while the group to the north, less than 100 feet away, sustained all the damage. This group consisted of six buildings which were only in the early stages of erection, and for the most part, had merely the steel structural work in place. This had the effect of a monster network, which, it would be supposed, the wind would pass through freely and without damage. Certainly it would be expected that the damage would be equal all along the line, but the result was far different. The order of the buildings from west to east was: Warehouse No. 1, warehouse No. 2, the stemmery, the powerhouse, the licorice factory and offices. The tornado passed the first half of warehouse No. 1, demolished the second half, overlooked No. 2, destroyed the stemmery, killing a number of men, neglected the powerhouse, and destroyed the licorice factory.

The damage was about about $150,000, and the Secretary of the company issued this statement the following day:

"The new plant of thirteen buildings in process of construction for the Liggett & Myers Tobacco Company has suffered seriously from the cyclone. The stemmery building, one-half completed, is a total wreck. The rehandling building, or warehouse, finished, is not badly damaged. Part of the roof and the cornice of the licorice factory have been torn away, but the loss is not serious. The east end of the box factory and the office building are badly wrecked.

"In relation to Thomas Clark & Sons, as contractors, they are required to furnish the building material and construct the plant entire. The company's contract with them calls for the delivery of the completed plant September 15 next. Of course, the contractors are the losers from the result of the cyclone on the buildings, but it is fair to presume that they are secured by bondsmen from sub-contractors in the same way that the company is secured by responsible bondsmen from them. If that is the case the loss will not fall heavily on any one of the contractors.

"Thomas Clark & Sons contracted to build the plant for $1,000,000. The company put in its own foundations at a cost of $250,000. Its officers feel confident that the contractors will be able to fulfill their agreement with it to the letter."

The site of this immense plant has Lawrence street as its eastern boundary and Tower Grove avenue as its western boundary. It has a frontage of 2,400 feet on Park avenue and a like frontage on Folsom avenue. The buildings are of iron and brick, and they cover nearly the whole of the site, while they are planned to range from four to six stories in height.

It was in one of these immense unfinished structures— the warehouse—that death came in a most sickening manner to workmen. The iron-work was four stories high, and beneath it was a basement eight feet deep. The workmen were seeking shelter in this basement when the ponderous iron-work, with its massive steel posts and girders, toppled before the violent cyclone and crashed down upon them. Nine-tenths of the victims were St. Louisans.

It was at one time believed the number of deaths here alone was in the hundreds, as 500 men were employed. Most of them had just left the building when it fell, and were thus providentially saved.

CHURCHES WRECKED.

The total loss of church property in St. Louis from the tornado amounted to over $400,000, distributed as follows: Roman Catholic, $225,000; German Evangelical, $55,000; Episcopal, $35,000; Methodist Episcopal South, $25,000; Baptist, $20,000; Presbyterian, $19,000; Unitarian, $10,000; German Evangelical, $10,000; Christian, $8,000. It is impossible to arrive at the exact loss, because none of the churches partially destroyed have been examined by the architects to ascertain what portion of them can be used in reconstruction. Conservative officers of the churches have estimated the amount of destroyed and damaged churches, twenty-four in number, in various amounts, which round out the sum total heretofore given. Of the number of churches given above, seven are complete wrecks. Of the Catholic Churches, the Annunciation and St. John of Nepomuk were totally destroyed. The Lafayette Park Baptist, Mt. Calvary Episcopal, Memorial Methodist South, McCausland Avenue Presbyterian and Compton Heights Christian Churches shared the same fate.

It will be some time until arrangements are perfected to rebuild some of the churches, owing to the great number of homeless families, made destitute by the same tornado that razed the churches. In the meantime other churches more fortunate, have been tendered the stricken congregations in which to hold religious services temporarily.

The following Sunday the various Catholic parishes bereft of religious homes were invited to masses in St.

NATIONAL HOTEL, EAST ST. LOUIS.

COURT HOUSE, EAST ST. LOUIS.

THE GREAT CYCLONE.   169

Vincent de Paul's Church, Ninth street and Park avenue, whose auditorium suffered but little injury. Its environs, however, and some of its "frills" were badly wrecked. Mass was celebrated hourly from 5 to 11 in the morning. The Lafayette Park Baptist congregation held services in the German Y. M. C. A. Hall, 1800 South Tenth street. The Lafayette Park Methodist congregation worshiped in a large tent on the church lawn, facing Park avenue. The Presbyterian Church, just around the corner, worshiped in the basement of their damaged church and The Church of the Unity congregation met in its Sunday-school room. The McCausland Avenue Presbyterian people held their meetings in the Grace Chapel, at Goodfellow and Theodosia avenues.

The German congregations were invited to unite with other German churches most convenient for them to attend. Other homeless congregations were similarly provided for by churches of their respective denominations.

A QUARTER OF A MILLION.

One corporation lost $288,000 by the storm, of which only $45,000 was protected by cyclone insurance. This was the United Elevator Company, organized in 1889, with a capital of $2,085,000, and a bonded indebtedness of $1,215,000. In May, 1895, a shortage of wheat was discovered in the stock in the elevators. This amounted to $250,000 bushels of wheat and entailed a loss of $150,000 to $200,000, the exact figures not being obtainable. In the summer of 1895 it was discovered that a large amount of the corn in stock had heated, and this entailed a further loss of $125,000. In order to meet these losses and to provide money for operating expenses it was necessary to place a blanket mortgage of $485,000 on the property.

The business during the winter was disappointing. The wheat crop of 1895 was a failure in the territory contiguous to St. Louis. The corn crop of 1895 was diverted from this city, being shipped by rail to Galveston and New Orleans for export. The result has been that for 12 months past the elevators have had practically no business and were operated at a great loss.

A month ago a movement was set on foot to fund two years' coupons on the first and second mortgage bonds for five years. All the bondholders were not agreed to this. A majority were. Several large bondholders, notably Messrs Kehlor and Haarstick, would not agree to the scheme unless $25,000 was expended on the East St. Louis Elevators to put it in repair.

STRUCK BY THE TORNADO.

The matter was still under consideration when the tornado struck the city and forever sealed the doom of the St. Louis United Elevator Company. Every elevator in the system was damaged by the storm save the one at Venice, Ill.

Unfortunate as the company has been it was the most unfortunate as to the loss of any corporation in the city. Its individual loss is the greatest in the city. The wind wrecked its elevators and its prospects at the same time.

The total loss on its property is given at $288,000. On this it had but $45,000 of tornado insurance, leaving a loss of $243,000 above insurance. The loss is divided among the elevators as follows:

Valley (totally destroyed)	$130,000
St. Louis (partially destroyed)	25,000
Merchants' (partially destroyed)	10,000
East St. Louis (partially destroyed)	15,000
Advance (partially destroyed)	30,000
Union (partially destroyed)	25.000
Central A (partially destroyed)	5,000
Union Depot (partially destroyed)	8,000
Total	$288,000

AN APPALLING LOSS.

This loss to the company is appalling. It is already bonded for $1,700,000. This is as much as the properties are considered to be worth, although the capital stock is $2,685,000. It would cost $1,250,000 to rebuild the elevators entire, but most of them are old, and two-thirds of that amount would represent the value of the buildings before the storm. Of the bonds $1,215,000 are firsts and $485,000 seconds.

It is doubtful if the Valley or Advance elevators will admit of repairs or even reconstruction. This being the case and the management, finding that it had not enough funds on hand to care for the grain in store, to say nothing of the repairs to the damaged property, at once notified all holders of grain in its houses that they must make provisions for the care of the grain as the elevator company was unable to do so. This condition of affairs, together with the fact that $4,110 interest was due on the second mortgage bonds, $14,550 on the second mortgage bonds just after the storm, $22,665 on first mortgage bonds July 1, and $5,400 August 1, made a receivership inevitable. Two days after the storm the C. B. & Q. Railway Company, which owns 1,900 shares of stock, $97,500 in first mortgage bonds and $35,000 in second mortgage bonds, with R. P. Tansey, who also is a stockholder and bond owner, applied for a receiver. As both are non-residents of Missouri they filed their petition in the United States Court. Judge Adams heard the matter in chambers, granted the petition and appointed ex-Governor D. R. Francis receiver.

DAMAGE EAST OF BROADWAY.

The steel structure that supports the tracks of the Terminal Association acted as a sort of protector to the fronts of the buildings that face the river from Washington avenue to Poplar street. It served, also, as a convenient rack for the hurricane to hang sundry and divers articles, brought over from Illinois, upon. These varied from a tattered shirt, which waved in the wind at the foot of Locust street, to a heavy beam weighing several hundred pounds, which had to be removed from the tracks before travel could proceed.

Most of the houses along this locality are very old and it is a wonder that more were not destroyed. They are mostly occupied by levee saloons and chandler stores on the ground floors and the upper stories are given over to cheap lodging and tenement houses. To the occupants of those houses it was a night of terror.

AGRICULTURAL WAREHOUSE ON CARROLL STREET.

MOUNT CALVARY CHURCH, LAFAYETTE AND JEFFERSON AVENUES.

THE GREAT CYCLONE. 175

After the fury of the storm had passed, their fear refused to be allayed and they huddled in groups until the day broke. Many were afraid to remain in the houses lest there should be a repetition of the violent wind and spent the night wandering along the levee front. With daylight, however, their courage returned and they thronged the levee laughing and discussing the events of the night as if nothing had occurred to upset their equilibrium, and morbidly curious, now that the danger was past, to gain all the gruesome details of the disaster. Here and there could be seen the tear-stained countenance of one who had lost a dear one in the course of the night's horrible events, but the majority of that motley crowd seemed unconcerned and even gay.

The cyclone that caused the damage along the river front evidently crossed the river diagonally from the northwest. Its violence is noticeable all the way from Washington avenue to Lesperance street, where it seems to have veered westward. Along this entire distance the levee is thickly strewn with rubbish and debris. The wind was so violent that the waters of the Mississippi were at numerous places carried up into the lower stories of the houses fronting on the river. The steel posts that support the terminal tracks show marks that insure that the water must have flowed several feet deep over portions of the levee that are dry this morning.

A number of sheds were standing near the elevator. In one of them was a horse and buggy belonging to an employe. Neither shed, horse nor buggy has been seen since the storm. All have completely disappeared, leaving not a trace. The body of a horse, destitute of harness, was found half a mile south the following morning, but has not been identified as the one carried from the elevator yard.

As an illustration of the freaks played by the wind, a large wagon loaded with lumber, which an ordinary gust would be sufficient to overturn, was left standing in the yard uninjured, while the horses hitched to it were torn loose and whisked away.

The sides of the elevator were covered with heavy iron sheeting, strongly bolted to the beams. All this was stripped off as cleanly as if done by human hands. The lofty brick chimney withstood the wind's assault unscathed.

The large buildings occupied by the Campbell Glass and Paint Company and the Geo. P. Plant Milling Company were also badly damaged. It is impossible at this time to estimate the damage in dollars and cents.

A PANIC AMONG FACTORY GIRLS.

The factory of the Bemis Bros. Bag Co., at the foot of Poplar street, had one side completely blown out. Several hundred girls who work there were panic stricken and rushed in all directions. Quite a number are believed to have been injured by flying timbers. The next morning the girls assembled about the factory in droves and discussed innumerable hair-breadth escapes they had experienced. Then they went sight-seeing with evident enjoyment, regardless of the fact that they will be thrown out of employment.

The naval store and department of the Waters-Pierce Oil Company was completely destroyed. Malachy McDonald, the superintendent, was struck by a piece of falling roof and instantly killed. The other employes escaped with cuts and bruises. Bob Newman, an employe of the company, is congratulating himself on his good fortune. He left his office about a quarter to 5 o'clock, and had just reached Broadway and Olive street when the storm broke. The next morning he found his office a mass of ruins. McDonald was killed within a few feet of it.

DEVASTATION'S AWFUL TRIUMPH.

At Levee and Pine street seemed to mark the northern limit of the real hurricane. At that point several buildings were partially wrecked. Commercial street between Pine and Chestnut was so obstructed by debris as to be almost impassable. Along this block almost every house was unroofed and two or three of the taller ones have lost the upper stories. The building occupied by the St. Louis Commission Company was entirely wrecked and its contents scattered over the neighborhood. Sacks of wool were carried blocks away.

At the Levee and Market street several small buildings were demolished. They were occupied by families, most of the inhabitants being more or less injured. The rear of the large factory of the Bridge Beach Company was wrecked, the roof and two upper stories being carried away.

Devastation on every side greeted the eye of a person

walking south on the Levee, but not until Poplar street was passed could the full violence of the storm be appreciated. Between there and Carroll street there was not a whole house standing. At the foot of Cedar street a two-story building was completely razed to the ground. There is doubt as to whether it was occupied or not. People living in the neighborhood say that it had no regular inhabitants, but almost every night a number of refugees sought shelter there.

The immense Nedderhut warehouse was partially wrecked, but no one is thought to have been killed there.

The wind played havoc in the Iron Mountain yards. Freight cars were blown about like they were baby carriages, overturned and wrecked. Whole lines of cars are so covered with debris as to be almost invisible. A dozen sheds and small buildings that served various purposes about the yards have completely disappeared. The main building is shattered, but still stands. The most appalling sight along the demolished river front was the wreck of the United Elevator at the foot of Chouteau avenue. This immense structure offered a fair mark for the cyclone, and besides lay almost directly in the center of its path. From the main building the four upper stories were torn. So clean was the wind's work that at a distance it looked as if they might have been sawed off. For hundreds of feet on every side masses of broken and splintered lumber were strewn in a way to make approach difficult. The river house was also demolished and the network of trams that connected it with the elevator proper obliterated,

FRONT VIEW OF ANCHOR HALL.

SIDE VIEW OF HODGEN SCHOOL.

THE GREAT CYCLONE. 181

Half a dozen rushed into the office, a little frame shanty on the north side of the elevator. Although it was the frailest structure of the lot it was the only one that remained intact. The terror-stricken men barred the door and then clung to each other in momentary apprehension of a terrible death. Timbers and flying articles of all kinds dashed against the shanty and smashed the windows, but fortunate none of the men were struck. Several times it seemed like the house would be carried away intact, but an immense safe acted as an anchor and held it down. C. E. Allen, the day watchman, started to run across to the river house, but his companions saw a piece of flying timber strike him and he was buried in the ruins of falling buildings. Quite a number of men and boys were observed to be fishing from the river house just before the storm, and are supposed to be buried or to have been washed away by the waves.

STATION TORN UP.

The small station at the junction of King's highway and the Oak Hill railroad was completely torn up and thrown to the ground. The station master, Pat Higgins, was in the building at the time and was buried beneath the debris. By the prompt assistance of witnesses of the disaster he was finally extricated. His head was badly cut and he was bruised all over his body. His injuries are not serious.

The storm did much damage to Charles Shewer's saloon at King's highway and New Manchester road. The stable was unroofed and the horses narrowly escaped being killed.

Mrs. Bredemeyer furnished one of the few humorous gleams across the dark tale of suffering and disaster which the storm has left in its wake. Mrs. Bredemeyer went out to the cow shed to get some bran for her cow's supper. She did not find the cow shed and it was not until she had gone around like Diogenes with a lantern that she discovered the shed up in her attic. She did not attempt to explain the trick the wind had played upon her.

IN NORTH ST. LOUIS.

At Broadway and Carr street the roof of a three-story brick building containing stores and flats blew off, a large section of the roof striking grip car No. 4 of the Broadway cable. The car was wrecked, the passengers escaping with slight injuries. A portion of the same roof fell upon the Rialto Livery Stable on Carr street, burying two of the stable employes and several horses. The men were taken out uninjured.

The Niedringhaus Memorial Hall, at Seventh and Cass avenue, was unroofed and considerable damage done the library and other contents of the building.

The roofs of two of the buildings of the St. Louis Stamping Company, at Second and Collins streets, were torn off and eight large smoke stacks blown down.

The driver of a bakery wagon belonging to B. Colombano, 1411 Pine street, was probably fatally hurt while passing the corner of Sixth and Carr streets. The wagon was completely demolished by a falling wall, and the driver removed to the City Hospital with a broken leg and serious wounds about the head.

The wall of B. Tegetoff's saloon, at Seventh and O'Fallon streets, fell on a Cass avenue motor car, No. 55, and a number of passengers were hurt, none seriously.

The steeple of St. Patrick's Church, Sixth and Biddle streets, was blown down and the adjoining residence of Father McCaffrey unroofed. A squad of police from the Third District Station procured tarpaulins and covered the furnishings of the residence before great damage was done. The roof of the parochial school was also blown off. Gilbert Chapman's cigar store, at the bridge entrance, Third and Washington avenue, was completely leveled. The proprietor and five unknown men in the building escaped injury. It was a small structure, valued at about $600.

PANIC AT THE RACES.

At the Fair Grounds races, 3000 persons received a fright that they will not soon recover from. The roof of the grand stand was blown off and completely demolished, a portion of it striking a horse hitched to a milk wagon on the Natural Bridge road. The horse was killed instantly. No lives were lost, and that such is the fact is indeed miraculous. The rain probably saved many persons from being crushed to death by falling portions of the room. Whenever the visitors at the Fair Grounds are overtaken by a heavy rain all hands either seek the basement or betting ring as places for shelter. When the storm broke the portion of the crowd not in the betting ring at once turned down into the basement. There they escaped being injured.

The crashing of the roof on the ground was like the explosion of a hundred cannons and together with the thunder and lightning sent fear to the hearts of every person on the grounds. Every one thought of the cyclone and its terrible ravages and pandemonium reigned for fully fifteen minutes. Women became hysterical and ran around the basement like mad while strong men were terror-stricken and speechless. Only the presence of mind of a few kept the crowd from surging out of the doors leading to the north walk, where certain death from flying timbers awaited them.

In the betting ring, where many speculators were in line trying to cash on the fifth race, which had just been run, and others waiting for the odds to be posted on the sixth event, the greatest confusion prevailed when the storm lashed forth in all its fury. The betting shed stood well the test of the wind and the only damage done was the blowing away of the awnings that surrounded the affair. As the wind continued to blow, several hundred persons who fancied that it was only a matter of a few moments when every building on the grounds would be razed to the earth, ran as fast as their legs would carry them across the track to the center field. There they stood amid the thunder and lightning watching the wind spend its fury.

Following the wind and electric storms the rain came down in torrents. The people in the field were soon drenched to the skin and with feelings of the man who jumped into the river with all his clothes on.

A horse and buggy belonging to the Sano Chemical

D. M. OSBORNE & CO.'S BUILDING.

RESIDENCE ON PARK AVE.

Company was standing on the north road when the storm broke loose. A large portion of the roof fell down upon them, and not a person who saw the horse struck believed he was alive. When the storm had subsided scores of willing hands were at work with axes and hatchets attempting to extricate the animal from the debris he was under. After fifteen minutes' hard labor the roof was raised, and the horse pulled out from his position—alive, and not a scratch on him. This is an illustration of the many miraculous escapes that both human beings and animals had in the course of the afternoon. The force with which the falling portions of the roof struck the ground was something tremendous, as evidenced by the split telegraph poles which happened to be struck on the Natural Bridge road. The storm put an end to the cashing of winning tickets on the fifth race, for speculators and bookmakers flew in all directions. Many made their escape to the hedge along the outside fence, where they sat down in the mud and slush awaiting their fate.

No further damage was done at the fair grounds other than the blowing down of a number of trees. The Fair Association will sustain a loss of about $1,500. Secretary Aull stated that it would cost no more than that amount to put on a new roof to the grand stand. A wrecking company was put to work next morning and the debris speedily removed. Racing was continued the next day, only the lower part of the grand stand being used by the visitors.

RESCUED FROM A WRECKED SALOON BUILDING.

The great Cupples buildings withstood the storm well, but hundreds of dollars' worth of plate glass was destroyed. All the windows in the south and eastern ends of the buildings were blown in and the upper floors flooded with the heavy rain. Much of the goods stored near the windows were damaged by the rain. No one was injured. Across the street, the small frame and brick buildings suffered. Nearly all of them were unroofed and many of them had their second stories blown in. The saloon of Peter Nalty, at 614 South Seventh street, was completely demolished. Nalty, his wife and nine customers were buried in the debris. All of them were rescued alive and sent to the Dispensary for treatment. Nalty lived over the saloon, and lost not only his business, but his home and its furnishings as well. Curley's lodging-house, at 628 South Seventh street, was demolished, but all the occupants escaped and most of them helped dig out the unfortunates buried in the wreck of Nalty's saloon.

The Missouri Pacific headquarters building was badly damaged, and the freight sheds at Seventh and Valentine streets were completely unroofed. The street was filled with broken wires and poles from Market street south, so that travel was almost impossible.

The large furniture store of Wm. Ottenad, at Soulard and Broadway, was completely destroyed. Ottenad, his wife and a clerk and a driver were buried in the wreck, which is possibly the most complete one in the city. The building was a four-story one and nothing but a heap of broken timber and dismantled stone and brick marks the spot.

The Soulard Market building was unroofed and a portion of the building was blown down. The old St. Vincent's Insane Asylum was unroofed and flooded and the great brick wall was blown down. All along Lafayette avenue houses were unroofed and moved from their foundations. The streets were filled with trees and broken wires.

DOWNTOWN WRECKS.

The feed store of Arnold Fuchs, at Ninth and Clark avenue, was almost demolished, the roof and front of the western portion having been blown away.

Morrissey's saloon, on Eleventh and Clark avenue, had the front and eastern walls blown out. A two-story frame house and a brick carpenter shop adjoining it on the east were literally wiped out of existence, and the contents of Heinrichshofen & Lawrence's lumber yard, on Tenth and Clark avenue, were scattered as far east as Seventh street.

Enno Sanders' bottling factory, on Eleventh, near Walnut street, was almost totally demolished, the northern front portion, together with the roof, going by the board.

The roof of the Sumner High School, on Eleventh and Spruce streets, took a vacation, joining the debris from the roof and walls of the jail, which blew out onto Spruce street. The roof of Tom McDermott's saloon, on Eleventh and Chestnut streets, fell off and tried to get into the front door.

All the windows in the front of the Hagan Theater were blown out and a portion of the roof demolished.

The heavy iron roof of the Merchants' Exchange building was lifted off in three sections and deposited on Third and Chestnut streets. A portion of it caught on the wires on Chestnut street, where it hung down like a huge tarpaulin.

The roof and upper front wall of the "Amerika" building, opposite the Exchange building, was blown down into Third street.

Perhaps the most unique freak played by the wind was observed on the tower of the McLean building on Fourth and Market streets. The observatory of the tower was filled with chairs, signs, awnings and other debris that had been caught up by the wind from the street, or blown out of adjacent offices and whirled up a distance of eighty feet in the air, cathing on the projections of the tower, where they remained, a most practical illustration of the fantastic power of the tornado.

The entire roof of the mammoth warehouse of Martin Lammert, on Second and Walnut streets, was uplifted and scattered over several acres of the vicinity.

On Sixth and Chestnut streets the wind caught a heavy four-horse truck and blew it 100 feet up the street without overturning it.

At the stables of the American Express Company, on Eleventh and Walnut streets, half a dozen of the heavy

WRECKED HOME ON OREGON AVE.

SAMPLE OF THE STORM'S RAGE ON PARK AVE.

express wagons were overturned by the wind like so many toy carts.

On Fourth and Market streets, within a space of 200 feet, no less than fourteen overturned wagons were counted during the height of the storm. Four horses were blown out of the harness that attached them to the vehicles, and all found refuge in a butcher shop.

On Twelfth and Clark avenue an unknown individual deserted the horse and buggy he was driving. The wind struck the rig, whirled it around three or four times and blew the buggy clear of the horse, carrying it up Clark avenue to Eighteenth street, where it was demolished. The horse was blown up against a telegraph pole and stunned.

A man who was standing in front of Hoefner's saloon, on Twelfth and Clark avenue, was caught up by the wind and carried through the glass door.

CHAMBER OF COMMERCE UNROOFED.

The roof of the Merchants' Exchange was ripped from the fastenings and whipped about like a rag in the wind. Great sheets of tin were hurled up in the air and twisted into shapeless masses. There was not a square foot of the roof left smooth and when the storm was over the roofing was lying on the four streets surrounding the building.

A piece of the roof, 50 feet long, on the Pine street side, was lifted high up in the air and then carried downwards towards White's restaurant, on the north side. The strings of wires running up the street were all that saved the building, for one end of the sheet caught on the wires and it hung downward, reaching to the granite. While this was going on the same scene was being enacted on Chestnut street. Two pieces of the roof flew like handkerchiefs across the narrow thoroughfare, and again the wires saved the building. One hung down in front of the Booth Packing Company's store and the other rested on its center on the top of a telephone pole in front of the American Tent and Awning Company. These swung to and fro in the wind and threatened to fall at every moment.

On Third street the surface was strewn with brick, bits of roofing, cornice and wire. A large piece of the cornice was torn from the gable over the main entrance and struck one of the ornamental lamp posts near the steps. The iron post snapped like a reed and fell with the other debris. Some of the sheet metal roofing was blown as far as Olive street and was tangled with fallen wires.

The interior of the building presented a scene of utter desolation. The protecting roof being gone the rain came through the plastered ceiling in a perfect deluge, and the water on the floor was ankle deep. Throughout the entire structure the water poured as though no covering was over it at all. Every office from the cellar up was thoroughly drenched, and an umbrella was necessary while walking through them.

In the main hall workmen were busy all night sweeping out the water, but it was like trying to keep back the sea with a broom. There were no electric or gas lights in the place, and a hundred candles, with sample pans for improvised candlesticks, were stuck around to give light for the workers. The water simply rained in through the plastering, and the magnificent paintings on the ceiling were utterly ruined. So soaked was the plastering that it threatened to fall at any moment and bury the laborers who were sweeping the water out, but they worked all unmindful of their danger.

The pit was indeed pitiful. Water stood a foot deep in the ring, and reached above the lower step.

In the basement is an office which was kept busy all night. It is the supply department of the Bell Telephone Company, and the linemen were in and out at every moment, getting material to repair the damage to the wires. They, too, had to wade in water, and the clerks sat on high stools to issue the supplies.

SKYSCRAPERS ESCAPED.

It has been said that a lofty office building is dangerous in a storm. This did not prove the case with the St. Louis high buildings. In not a single instance was the slightest damage done to one of them, and tenants in the top stories felt no more inconvenience during the passage of the storm than if they had been on the first floor of a one-story structure.

A report was circulated and even telegraphed to other cities that the roof of the Rialto building, a ten-story office building, was blown away. The Rialto sustained no damage.

At the Union Trust building no damage was done and

the building did not even sway during the height of the storm.

The half finished Chemical building was open to the elements, none of the windows being in, but the damage to it did not amount to $10.

Other big buildings that went through without loss were the Commercial, the Laclede, the Columbia, the New Century, the Odd Fellows, the Security, the Burlington and others.

All of these dwellings are of what is known as steel construction. They are built with interlaced and riveted steel beams and joists, and are as solid as rock.

Their roofs, the portion of the building that would be expected to suffer in a wind storm, are as safe as the body of the building, as they are part of it, being built with it. To blow off the roof would mean to break the interlocking of the entire structure and great damage would result.

In the construction of these skeleton buildings the architects made specific calculations for resistance to a wind having a velocity of from seventy to ninety miles per hour.

The effect of a tornado on a tall steel skeleton building has never been tested until the present time. Since the construction of these buildings was begun in large cities a severe storm has never visited any of them.

AT THE COURT-HOUSE.

The storm struck the Court House at 5:05. The office usually closes at 5 o'clock, but two suits came in then and

INTERIOR VIEW OF THE OTTENAD FURNITURE CO.'S BUILDING.

WRECKED TREES IN LAFAYETTE PARK.

THE GREAT CYCLONE. 199

the filing took about five minutes. While the reporters were reading the papers the storm broke. For two or three minutes the electric lights had faded and revived and then faded again, and this indicated an unusual electrical disturbance, and finally expired altogether. Then the storm fell in its fury. Windows in the dome were ripped out, and gusts of wind shrieked through the structure and made dismal howls. The crash of falling glass was followed by flooding rain. Then the wind swooped down with a final supreme effort, but no further damage was done the old building.

Every electric light in the building was extinguished. This would have been disastrous for the Fire Alarm Telegraph Service but for the fact that the office had kept its gas in commission in anticipation of such an emergency.

AT THE CUSTOM HOUSE.

Several hundred dollars damage was done the Federal Building, the bulk of it in the Weather Bureau. It rained through the roof up in the tower just about as hard as it rained on the outside aud nearly ever window in the observatory was broken. For a time Observer Frankenfield was apprehensive that his valuable instruments would be damaged; indeed, it looked, when the storm was at its height, as though the entire tower and all its contents would take a sail over the roof of the Chemical Building.

SCENES AT THE HOTELS.

The Planters' Hotel lost several hundred panes of glass. The most damage was to the eleventh story, occupied by the help. The night force were, for the most part, asleep when the storm came on. Some were awakened by the crashing of their windows, and others by the rain that blew and poured in on them. Two or three slept through it all. Every room on the west side of the top story was damaged and a number also on the opposite side. Inside transoms were blown in as well as outside windows. Two or three of the girls were cut with glass, but nobody was seriously hurt. The damage on the floors below was mostly done by rain that poured in through windows left open. The glass was, except in a few cases, sufficiently thick to resist the wind. On every floor two or three windows were blown in. The rest withstood the force of the storm. Where windows had been left open carpets, bedding, and in a few instances the clothing of guests were soaked. Several transoms on the opposite side of the room, connecting with the hallway were smashed by the wind that blew through the room. Many of the girls employed at the hotel, and some of the men, too, thought their last hour had come. One man in the laundry is reported to have fainted from fright. Considerable excitement prevailed also among the guests. The parlor floor and corridors on the office floor were thronged during the storm.

The ladies' ordinary at the Southern Hotel and the kitchen were badly damaged. Several holes were torn in the roof over the ordinary, letting the rain pour in in torrents. The kitchen range was partially demolished, and the skylight overhead wrecked. Hot meals were suspended for dinner at the Southern that night. The chef was cut in the head and back with glass, and four others were injured. Dr. Powers, the hotel physician and surgeon, was kept busy nearly an hour patching up wounds. One of the injured was a woman, the others being men. The skylight in the roof of the hotel over the rotunda was damaged enough to flood the floor of the rotunda. Manager Lewis was one of the busiest men in St. Louis during and for several hours after the storm. No hotel of magnitude was seriously injured, and the new Convention Auditorium was so slightly damaged that it was repaired in three days. Fortunately it was just outside the storm area proper.

DANGER AT THE JAIL.

At the Four Courts the most fearful excitement prevailed. The woman's corner of the jail was blown down and bedlam turned loose. The first known of the storm was the violent shaking of the building, which is one of the strongest and safest institutions in the city. The wind played havoc with the windows. Crash after crash would bring down showers of glass. The court was just about closed, when several bricks from the roofs of a neighboring building came flying through the window. The women among the witnesses screamed, and burly men pushed and trampled their inferiors under their feet. The mob was so excited that the clerks voice could not be heard as he cried out for the people to take their time. But it was useless, as the crowd surged forward and out of the doors. Luckily, no one was seriously injured.

A terrible whistling noise, followed by a series of frightful screams, came out of the jail. The prisoners, penned in the cells, cried to be let out.

Deputy Sheriff Wagner ran into the bull ring and ordered silence, but they cursed him and kicked against the door. A hurried conclave was held, and it was decided to let them out. About fifteen cells were opened, and the prisoners flocked out. Other cells were being opened, when a terrific crash of thunder came.

Again the prisoners who were still confined screamed to be released, and the lucky ones who were already out gathered around the winding stairway where the guards were assembled. Again and again the lightning flashed, and with a horrible crash the corner of the women's division of the building gave way. Then from the interior of the jail came shriek after shriek. The prisoners ran up and down the winding stairs and tried to force the guards to let them out, vowing that they would make no effort to escape.

They became so wildly excited that Deputy Jailer Wagner ordered them again locked up in their cells. To this they objected, and stood the guards off. The riot bell was sounded, and in a minute a score of detectives and policemen ran into the jail. They tried to force the prisoners back, but it was useless. Then the officers drew their revolvers, and with clubs lifted aloft threatened to kill the men if they did not return to their cells. This frightened them, and they went back to their cells. Then a dozen officers were stationed at the corner of the building, which was blown down, to prevent the men from es-

M. M. BUCK & CO.'S WAREHOUSE, 12TH AND GRATIOT STREETS.

VIEW ON LAFAYETTE AVE.

caping. A milk-white pigeon fell down from the roof on the bricks below.

The top of the building suffered the worst. It shook so that the building fairly rocked. Matron Harris, who is on the fourth floor, came running down, frightened half to death. She was taken into Chief Desmond's office, which escaped the storm's wrath. In Chief Harrigan's office two window frames were blown into the room and the carpet was soaked with water. They were picked up by the wind and carried into the middle of the street. A man driving a butcher wagon attempted to drive down Clark avenue, but his horse took fright and upset the wagon. He was caught by the wagon and dragged about 25 feet. Some men who were standing in the doorway ran and picked him up and carried him downstairs. He was stripped of all his clothing and was bleeding from every part of his body.

HEARTBREAKING SPECTACLE.

The scene at the Morgue all night long defies description. All through the darkness, and in the daylight which followed, bodies were brought there. The usual marble slabs were displaced by pine coffins, as requiring less space.

The bodies were stretched upon boxes, and before long every inch of space except a narrow passageway was occupied.

The crowd in front of the Morgue increased constantly until it blocked the sidewalk and extended into the street. It was a weeping and wailing concourse. Pitiful tales were told by mothers of their missing children, by wives of husbands who had failed to return home.

The people were necessarily shut out from their dead. Had all who sought admission been granted it the Morgue would have been filled to overflowing and the proper disposition of the bodies that were arriving, almost constantly, impossible.

The crowd, which was almost a mob, surged and groaned with the horror discernible through the glass partition. Bodies of men and women, many battered beyond recognition, met the gaze of those fortunate enough to reach the glass screen.

Some idea of it was passed to those behind by the exclamations from those in front.

At 9 o'clock the crowd in front of the Morgue was so dense that additional policemen were necessary to preserve order. To permit traffic the crowd was forced off the pavement and kept lined up at the curbstone, with a passageway left in the middle of the street for dead wagons.

There were constant appeals from frantic people for admission. Some were stolid in their grief, others gave vent to most heartrending outcries.

For the few who recognized their loved ones in the array of mutilated dead hundreds were left in suspense by failure to find those for whom they were searching. The number of missing in St. Louis on Thursday must almost equal the number of dead, for the proportion of missing reported at the Morgue during the morning is necessarily small.

While the attaches of the Morgue were busy with a body, a woman almost crazed with grief burst through the Superintendent's office and into the room of the dead. Her cries resounded through the charnel house as she called aloud: "O, my sister, my sister!"

It was impossible to quiet the frantic woman. She rushed, as one distracted, from body to body. Her sister was not there. She was not satisfied. She continued to cry out in her grief. When finally quieted she said that her sister's name was Ida Sieger and she lived on Sidney street, near Twelfth street. Her condition was such that she could give no further information.

Fifteen minutes before three women, scarcely less frantic, had appeared searching for three young girls who left home before the tornado Wednesday afternoon and had not returned. They had started for a physician's office on South Broadway. They were Lena Lange, Edna Wetzel and Olga Wehrfritz, all living at $2906\frac{1}{2}$ Olive street. One of the women was the mother of Edna Wetzel. They were so overcome with grief that it was with difficulty they made any statement.

BRINGING IN THE DEAD.

Momentarily the gong of an approaching dead wagon would cause the crowd to part. The trips of the wagons with their mutilated human freight became more and more frequent.

There had been a slight lull shortly after daylight, until the searchers at work in the path the cyclone had marked could reach the victims of its fury.

As the work progressed the wagons with the dead drove up to the Morgue almost in procession. Each new arrival was greeted with gasps of horror by the crowd.

What had at first been a collection of bereaved and grief stricken men and women, had gradually grown to a mass of curiosity seekers.

At one time the crowd almost took the Morgue by storm. Several men attempted to force their way into the office of Superintendent Mack, but were pushed back and the door barred. The police had difficulty in controlling the almost frantic mass.

Those who were seeking their dead were admitted to the Morgue in twos and threes. Many obtained admission under false pretenses. Some found their dead; others left with sinking hearts, convinced that their missing were beneath ruins.

Within 12 hours of the disaster there were 56 bodies in the Morgue, many of them still unidentified.

CHURCH ON MISSISSIPPI AVE.

COLLINS' LIVERY STABLE.

IN EAST ST. LOUIS.

The tornado that wrought such havoc in St. Louis jumped the Mississippi and wrecked the western half of East St. Louis. More than 100 people were killed and three times that number injured. Two million dollars will not cover the damage done to property. The damage was done in a few minutes' time, and how any person in the path of the cyclone escaped is a mystery to all who passed over the devastated section. The wind struck the levee just north of the East St. Louis elevator, about 5:30. The wharf of the Wiggins Ferry was the first to suffer, and it was thrown far up on the levee.

Eye witnesses of the scene in East St. Louis describe it as the most terrible ever witnessed. Whole roofs of houses were flying through the air. Trees were uprooted, and every movable object was endowed with the power of locomotion by the hurricane. Those who possessed sufficient courage to look from their windows saw a mass of opaque air, which seemed to be whirling around in a circle and possessed of no definite object except to destroy as much property and take away as many lives as possible. When the sharp clap of thunder, immediately preceding the fiercest part of the storm, reverberated throughout the city, there were many who thought it was the crack of doom. They sought refuge in prayer and by hiding themselves in their closets. Still the wind blowed. Every gust seemed to gain encouragement from the one which preceded it. It was no respecter of persons. Roofs of modest dwellings and those of factories and apparently solid structures were blown away. It lasted perhaps fifteen minutes, but time is merely a measure of events and so many things happened in that short interval that there were many terror-stricken persons who believed the storm lasted for hours.

When the clouds rolled away and a clearer atmosphere enabled persons to see some distance from their homes the city bore an altered appearance. Many dwelling houses were splintered into kindling wood. Men and women ran hither and thither, calling aloud for surgeons and nurses to help them care for their dead and wounded. Mothers implored help to find their children. Husbands carried the mangled bodies of their wives out into the streets, where the cool rain seemed a balm that would mitigate their suffering. Comparative calm came as suddenly as the storm. The wind ceased to blow and for a time comparative cheerfulness prevailed. When, however, the awful destruction of the storm was realized a gloom spread over the town like a pall.

PRAYED FOR MERCY.

It was converted in those short twenty minutes into a city where lamentations and prayers for mercy were offered, even after the source of danger had gone. The work of rescue began as soon as the inhabitants, stupefied by terror, realized that they might be of service to others. The rooming-houses and hotels being in a downtown district were visited by the rescuers first. Nothing but the sight of mangled bodies and groans from those who were pinned down by timbers could enforce upon those who had escaped a comprehension of the terrible destructiveness of the storm.

Residents of the stricken city gathered in little knots on the street. They did not mind the drenching rain. Nothing except the wind, which left death in its path, possessed any terrors for them. They saw that the entire east approach to the Eads bridge had been carried away by the wind as if it had been so much tissue paper. On every side ruins were to be seen. For a time chaos reigned. No one knew exactly what to do.

The Court House and police headquarters had been blown away. Officers knew not where to find their chief, and while vaguely conscious that they ought to do something were at a loss how to proceed. In the course of time, however, temporary police headquarters were established. Drays and wagons were pressed into service as ambulances, and the work of rescuing the dying and removing the dead began. Several persons were found whose bodies had been transfixed by the sharp ends of huge timbers. Others lay moaning and groaning under the weight of a ton of timbers. Every courier that arrived at police headquarters had fresh tidings of calamity, until those who received the reports became inured to tales of terror, and received them as quietly as though they had been commonplace annals of every-day happenings. With the storm came a darkness that added to the terror.

Residents of St. Louis, fearing for the safety of friends and relatives and of property besieged the watchman who had been stationed at the bridge with requests to be allowed to cross. The passageway was filled with debris from the bridge overhead. The watchman had instructions to allow no one to pass. Two trains which had been blown over by the storm blocked the way of pedestrians. The burning buildings in East St. Louis cast a lurid glare over the rapidly rushing waters and the ruins of the town pictured in silhouette led many to believe the conflagration had destroyed the entire city. Those who did get over saw only a scene of wholesale destruction of life and property. East St. Louis and its ruins were one huge mausoleum, covering no one knew how many dead.

SEARCHING FOR THE DEAD.

The scene here was frightful, but it was nothing compared with the scene along the Levee, Broadway and the side streets of the Island. Every house had relatives and friends among the dead and injured. Husbands, fathers and brothers worked at night in the freight houses, on the boats and the cars, and as these could not be found the anguish of mothers, sisters and other relatives was heartrending. They would run from one temporary morgue to another, and in some instances push aside dead bodies in order to attempt to identify their relatives, and, when found, the cries would melt a heart of stone. A mother would behold the distorted features of an only son, while at another place some other member would find a missing father, brother or daughter.

A HOTEL'S SAD ROLE.

The Tuttle House, Third street, was like a hospital all night. In one room upstairs lay William Cogan, one of

MAUCHENHEIMER PLACE, 7TH AND RUTGER STREETS.
(FIFTEEN PERSONS WERE KILLED HERE.)

RESIDENCE OF JOHN ENDRES, PARK AND MISSISSIPPI AVENUES.

the best known members of the Brotherhood of Locomotive Engineers. He runs on the Baltimore and Ohio Southwestern, and was to have taken out the 3 o'clock fast mail train on that road. He was in a barber shop at Third street and Missouri avenue when the storm came up. The building was completely wrecked and the half dozen people inside buried out of sight. Mr. Cogan worked his way out without assistance. and did not think he had been seriously injured. He suffered intensely from pains in his back and sides.

H. K. Vail, one of the boarders, had a broken arm in a sling. He was crossing the bridge when the storm came up. He hurried across and sheltered himself on the leeward side of the frame building, which blew over on him directly afterward.

One room at the Tuttle House was occupied by three young ladies. Maggie Herbert, of Washington, Ind., had a broken arm. "I was in the dining room of the Tremont House when the storm occurred," she said. "A portion of the roof blew off and half a dozen of us ran to get out of the way of flying timbers and bricks. We ran into the kitchen and back into the dining room, and finally the whole house fell in. Mrs. Hays, the landlady, and two or three others were killed and the rest of us badly hurt."

Miss Herbert's companions in the room were Florence Reilly, a telegraph operator, and Josie Gallen, a typewriter. Miss Reilly lives at 3672 Finney avenue, this city, and Miss Gallen at 3311 Chouteau avenue. They are both in the employ of the Illinois Central Railroad, and were at work in the company's freight office, on the Levee, when the storm struck the building. They went down with the structure, but each escaped with only a few bruises. They inquired anxiously about the effect of the storm in St. Louis, and particularly on Finney and Chouteau avenues.

C. E. Obrist, a boarder at the Tuttle House, got one of his arms badly cut by falling glass. He is an electrician.

Mrs. Dowd, one of the boarders, was missing. She went out for a ride on her bicycle just before the storm. Mrs. Dowd's husband is a government beef inspector at the National Stock Yards. He escaped without injury and hurried to the hotel to find that his wife was gone.

SCENES IN THE HOSPITALS.

A trip through the east side hospitals at midnight revealed terrible suffering. At St. Mary's the Sisters were treating fifty cyclone victims. Dr. McLean had charge of the surgical ward, and was assisted by Drs. Fairbrother, Le Haan and Thompson. The doctors had been busy since the storm sewing up and dressing wounds, The patients had sustained injuries of various kinds, but fortunately only a few were considered in a critical condition. Two or three died after being received at the hospital. Several had fractured skulls, and several others were injured internally. The reporter talked with a number of the patients.

Roy D. Moore, a Vandalia freight clerk, was found with his right arm broken between the elbow and wrist.

His home is on the Missouri side of the river, near the corner of Page and De Hodiamont avenues. He said he had been caught in the wreck of the Vandalia freight office, on the Levee, with about thirty others. He was pinned in near Mat Quirk, Joe Crean, Thos. Dougherty, Mr. Givens and Bob and Ed Bland. They were all taken out alive, and as to the twenty-five others he could not say what became of them.

On a cot next to Moore lay two children, Nancy and Albert Fierce. At their side sat an older sister and their mother, who had escaped with but a few bruises. The boy's shoulder and breast are badly injured, and Florence's right wrist is dislocated. Their home near the Crescent elevator, was blown over and demolished, the entire family being buried in the ruins.

The next cot west was occupied by a boy with his head in bandages. No one seemed to know who he was or where he had been brought from. The little fellow was asleep. Nearby was an unknown man with his head tied up. He had come from the operating room and was unconscious. One of the attendants said he was not expected to live long. He had been taken from the ruins of the Martel House.

In the same room was Frank Barr, who works in Nelson Morris' stock yards. He was at home, No. 7 Rock road, when the storm came up. His house was blown to pieces and he and his wife were buried in the ruins. He didn't know how seriously his wife was injured. His own injuries were not very serious, consisting of bruises and cuts.

In the hallway downstairs was John Malloy, who was caught in the ruins of the Air Line freight house and completely covered. He complains of pains in his side and left shoulder.

Thomas Dougherty, a Vandalia freight clerk, who lives at 1322 North Seventh street, St. Louis, escaped with a bruised hip and cuts on his head.

Al Tudrer was another storm victim found in the hallway. He comes from Hamilton, O., and came to East St. Louis a few days ago with Charles Kinney, of Indianapolis, in search of work. When they saw the storm coming they crawled into an empty box car. The car was blown over and the occupants badly bruised, one on the shoulder and head and the other on the hip.

In another ward Dr. E. Thompson was found sewing up a frightful gash in Patrick Trainer's head. The injured man was delirious and imagined that several men had him down and were beating him. A couple of men were holding him while the Doctor stitched up the wound.

In the same room was John McMahon, with his shoulder blade and rib broken and back injured. He came to East St. Louis from Alton a couple of days ago. He sought shelter, when the storm came up, in a little frame house near the Green Tree Hotel. The house was demolished and McMahon's clothes were blown from his body. He was cut and bruised all over, but no bones were broken.

Deamie Bender, who lives at 2609 Walnut street, this city, and works for the Big Four line, was found in that room with his feet injured. He went down with the freight office of that line, along with a number of others.

SCENE ON LAFAYETTE AVE.

OVERTURNED CARS ON 8TH AND PARK AVENUE.

THE BABIES TAKEN AND THE MOTHERS LEFT.

Joseph Burke occupied a cot in the hallway near the operating room. He works for the Air Line, and was in the act of closing the doors of the freight house when the building was blown down. Burke was injured internally and his left leg was broken.

Peter Harris, on the next cot, wanted the attendants to let his folks, at 1205 Liberty street, Kansas City, know that he had not been killed. Peter's left arm and one rib were broken, and he was otherwise injured, but will probably recover. He worked for the Illinois Central, but was injured at his boarding house on the levee.

On the second floor of the hospital was Mrs. Horace Trump, whose pathetic story would turn the coldest heart. Mrs. Trump was at the bedside of a sick sister, with her year-old baby in her arm, her 4-year-old daughter being near her on the floor, when she saw the storm; but it was impossible for her to get out of the room, so sudden was its approach. The storm in all its fury struck the house and the flying missiles struck her, knocking down the walls of the building over her and her two daughters, killing the baby and 4-year-old daughter instantly. Mrs. Trump was considerably bruised and hurt internally, and it is as if by a miracle that she escaped the fate which befell the children. It was fully four hours before she was rescued from the position and taken to the hospital.

Miss Minnie Evans was brought in suffering with a broken leg.

Fred Weide, a railroad man, hurt in the tumbling walls of the Tremont house, suffering with a double fracture of the right ankle, a dislocated shoulder and several broken ribs. He was in the parlor of the hotel with several others, saw the storm coming and thought there was no danger; that the building was strong enough to withstand its fury, and made no attempt to escape. He was caught in the collapse.

Wm. Murray was very seriously injured internally, as well as suffering from a broken arm and broken limb, also a dislocated hip.

Mary Hanet was very seriously injured about the head, and also internally.

Joe Duffy, a clerk of the Vandalia Railroad, suffered intense pain from a very painful injury. Mr. Duffy had the whole right side crushed in.

WHEN MORNING CAME.

When the dawn came it was possible to see the devastation wrought by the storm in East St. Louis. The sky was clear, and the beauty of the morning strikingly contrasted with the scene of desolation that was disclosed.

On the river bank, from Kehlor's mill on the south to the elevators on the north, not a house was standing. These huge structures and the cold storage company's plant were badly damaged. The river bank was lined with wrecks of boats.

With the river banks as the base, the entire triangle formed by what is called the Island, there is not a whole house standing. Even the Relay Depot had its corners broken and two huge roundhouses were shaved off below

THE GREAT CYCLONE. 225

the tops of the middle of the locomotives which stood within them.

One brick house stood without one wall, disclosing the interior and furniture exactly as the dwellers had left it. The pictures, beds, bureaus, washstands, chairs, tables and even the lamps on the tables were undisturbed. A room or two of several houses were left.

For the remainder either the walls and roofs of the frame houses were folded together like cardboards and lay flat on the ground or they were broken into kindling wood and were scattered to the four points of the compass. The brick houses were heaps of building materials.

It is marvelous that 10 persons escaped from buildings so completely wrecked.

The Island, on each side of the roadway, looked like a vast lumber and truck heap.

Freight cars were overturned or wrenched from their trucks and turned completely upside down.

Across the creek, although few houses escaped damage and many were demolished, the destruction was neither so general nor complete. But in every direction the eye rested on ruins. Walking the length of Missouri avenue and looking up and down the cross streets one would say that a third of the houses were wrecked and seven-tenths were damaged.

THE DEATH ROLL.

Identification of the dead was in many cases difficult, and the lists were still growing when this work went to press. Following is the latest revision of the sad roll-call in St. Louis, Mo.:

Anderson, Wm. F., 38, 606 South Seventh street.
Altus, Henry, 62, 3026 LaSalle street.
Anderson, Richard, 1029 North Seventh street.
Allen, Charles Edwards, 43, Platte City, Mo.
Alcornero, Charles, 19, address unknown.
Ahillern, John, 26, 2929 Missouri avenue.
Archambo, Alex, 45, City Hospital.
Bradshaw, Wallace, 20, 923 N. Jefferson avenue.
Bowler, Wm., 28, 328 Montrose avenue.
Benwall, Fred., 24, Jefferson and Shenandoah.
Boecklin, Ulrich, 48, 1309 South Seventh street.
Bergeest, John, 27, 2408 South Broadway.
Bohle, Fred., 64, 2807 Wisconsin avenue.
Benz, George, 22, Twentieth and Gratiot streets.
Bolm, Augusta, 63, 1706 Park avenue.
Bene, Sylvester, 5, 1418 Ohio avenue.
Carnero, Charles, 19, 1411 Papin.
Claypool, Cora, 67, 2641 Papin.
Claypool, Ethel, 10, 2641 Papin.
Cahill, Mrs. Mary, 70, Jefferson avenue and Papin.
Craddock, Martin, 36, 1211 Carr street.
Cheney, Emma L., 42, 1432 Mississippi avenue.
Crook, William, 33, 1611 Lafayette.
Crump, Catherine A., 47, 2711 Park avenue.
Creamer, Kate F., 20, 2643 Caroline street.
Craig, Charles, 817 Walnut.
Dieterich, Peter, 67, 1516 South Tenth street.
Duggan, Rose, 55, 1527 South Eighth street.
Dunn, Michael, 42, 1213 North Fifteenth street.
De Martini, Sophia, 16, 402 South Twelfth street.
Eyman, T. A., 47, 3129 South Jefferson avenue.
Enders, Charlotte, 40, 616 Rutger.
Elser, Joseph, 32, Gas Company stables.

A RUINED HOME.

EADS BRIDGE, EAST SIDE.

Fischer, Frank H., 6, 1944 Papin.
Friesecke, Mrs. Clara, 30, 609 Hickory.
Friesecke, Estella, 6, 609 Hickory.
Friesecke, Edna, 4, 609 Hickory.
Fiegler, Caspar, 40, 3313 Salina.
Gall, Julius, 26, 1507 Sulphur avenue.
Gearse, Julia, 64, 1306 South Third street.
Gibson, Henry, 38, 5526 Odell avenue.
Gray, Leroy, 5 months, 1523 Hickory.
Gegan, Henry P., 34, St. Louis House.
Gardner, Anna, 24, 1848 Menard.
Gardner, Emma, 1, 1848 Menard.
Goff, James, 29, 3715 Chouteau avenue.
Gallagher, Chas. L., 4, 2136 California avenue.
Gregory, Wm., 20, 1213 North Sixteenth street
Goodman, Adolph, 40, 1816 Linn street.
Hessel, John, 17, 1109 Emmett.
Hess Harry, 28, 1109 South Seventh.
Herbert, Geo., 22, Minneapolis, Minn.
Helix, Mrs. Melanie, 69, Seventh and Rutger
Howell, Mrs. Alice, 27, 714 Rutger.
Howell, John, 35, 714 Rutger.
Howell, Ida, 7, 714 Rutger.
Hassenfratz, Daniel, 71, 1014 Armstrong.
Horne, Isa, 26, 1432 Mississippi avenue.
Hickey, Maggie, 21, Vail place.
Hermann, Catherine A., 58, 3644 Castleman.
Hollerman, Taylor, colored, 38, 1401 Missouri.
Irvin, Thomas, 65, 2220 Chouteau avenue.
Jones, Richard, 35, 2809 Easton avenue.
Jacobs, Bertie, E., 21, 1510 Vail Place.
Jones, Thomas, Central Home of Rest.
Knoll, George, 55, 1801 South Thirteenth street.
Kuehling, Henry, 18, 2824 Oregon avenue.
Killian, Thomas, 40, 1303 South Seventh street.
Killian, Harry, 9, 1303 South Seventh street.
Killian, William, 7, 1303 South Seventh street.
Knobel, George W., 43, 1303 South Seventh street.
Keim, George, 72, 2111 Kosciusko.
Labar, John, 21, killed at 1819 Chouteau.
Loeblein, John, 49, 916 Barry.
Lanahan, William, 42, gas company's stables.

Lanahan, James, 26, 2005 Biddle.
Ligon, Mrs. J. J., 50, 2358 Park avenue.
Matz, Joseph, Belleville, Ill.
Mauchenheimer, Fred, 63, 1300 South Seventh.
Mauchenheimer, Mrs. Catherine, 56, 1300 South Seventh.
Maurer, Joseph A., 54, 1830 South Eighteenth street.
Miller, Joseph J., 30, 1510 Pennsylvania.
Moraghan, Thomas L., 24, 1653 South Jefferson avenue.
M'Donal, T. Malachi, 34, 2745 Clark avenue.
M'Given, Mary, 1, 2725 St. Vincent.
Nee, Chas., 35, 1503 Chestnut.
Oates, Thos., 55, 2230 O'Fallon.
Ottensmeir, Aug., 23, 3431 Gravois.
Ottenad, William, 41, 1700 South Broadway.
Osterman, Annie, City Hospital.
Platschek, William, 58, 614 Rutger.
Pomely, William, Second and Chouteau avenue.
Poppitz, Thomas, 20, 2501 South Broadway.
Proute, Katherine, 67, 1500 South Thirteenth.
Plank, William, 1848 Menard.
Rafferty, John, 23, 2223 Wash.
Ruebeck, Chas., 55, 1236 High street.
Rux, Matilda, 56, 716 Barry.
Rux, Tina, 17, 716 Barry.
Rohlfing, Anna, 23, 2122 Geyer avenue.
Reis, Theodore, 8, 2140 Chouteau avenue.
Rodriguez, Mrs., 40, City Hospital.
Sims, L. F., 37, 1707 Park avenue.
Selp, Edw., 26, Lami and Pestolozzi.
Sherbold, John, 53, 714 Julia.
Smith, Andrew, 4 months, 1523 Hickory.
Schmalenbach, Herman, 1700 South Broadway.
Schwerdtman, Chas., 29, 2648 St. Vincent avenue
Stephens, Thaddeus J., 23, 3728 North Market.
Steinberf, Christian R., 23, 2648 Allen avenue.
Schmidt, Chas., 37, 2633 Hickory.
Schueringer, Alex., 23.
Smith, Wm. Q., 82, Memorial Home.
Silva, Benj., 55, 3042 Locust.
Spillman, Martha B., 73, 2711 Park avenue.
Steinkoeter, Adam, 15 days, 1622 South Thirteenth.
Sudhoff, Chas., 39, 1805 Dolman.

Smith Samuel, Central Home of Rest.
Tandy, Chas. A., 3135 School street.
Trachter, Wm., 64, Third and Rutger.
Tainter, Chas., 3 months, 1522 Hickory.
Taylor, Wm., 45, 1401 Missouri avenue.
Tolbert Mary, 30, 213½ Market.
Teva, Anna, 1007 Allen avenue.
Vignette, Louise, 89, 1300 S. Seventh.
Vollmer, Gustav, 69, 1300 S. Seventh.
Winkler, William, 30, 2011 Virginia avenue.
Woods, William H., 29, 1428 S. Eighteenth.
Woodruff, Sarah B., 50, 2743 Ann avenue.
Wells, Fred, 2 years, 1728 S. Ninth.
Wells, Theresa G., 9 months, 1728 S. Ninth.
Wilson, Robert, 23, 4214 Cottage avenue.
Wills, Michael, 46, 1315 S. Sixth.
Wagner, John, 33, 919 Geyer avenue.
Weis, Max, 49, 3100 Magazine.
Weckerman, Anton, 40, 1938 Cherokee.
Zimmer, Ernst, 32, 4731 Greer avenue.
Zimmerly, Samuel, 56, 321 Russell avenue.
Zeic, Gregor, 2507 S. Second street.
Unknown white man at Morgue.

Among the list of names at first published as reported missing, the following have since turned up alive and unharmed:

Captain George Zeigler, of the City of Monroe.
Charles and Eddie Appel, 2904 Michigan.
Miss M. Paule, 2630 S. Fourteenth.
J. W. Johnson, Terminal Hotel.

Inquiry into the cases of George Hesse, reported missing from 1707 Jefferson avenue; Mollie Thurns, reported missing from Jefferson and Gravois, and Lizzie, a colored woman, reported missing from 2728 Russell avenue, disclosed that no such persons had been known in those respective neighborhoods, and the reports, consequently, lack verification.

All of these went down with the steamer J. J. Odill, just below the bridge.

Morris Fisher, Hardin, Ills.
Mrs. Gorgar, Hardin, Ills.
Sim Woods, clerk.
Louis Morris, cook.

The following river men, at first reported missing, were found to have miraculously escaped:

Jas. Flanagan, carpenter Vicksburg.
Jos. Jovin, diver.
Larry Dauer, pilot ferryboat Christy.
Henry Levy, fireman ferryboat.
G. A. A. Simons, second clerk Odill.
Jas. Boland, captain Pittsburg.
Geo. Townsend, captain Odill.

Among those known to be dead are:

Barber, name unknown, buried under ruins of Anchor Hall.
Colored driver for Polar Wave Ice Company, at Union Depot power-house.
Unknown teamster at same place.
Unknown Bohemian woman, in ruins at Thirteenth and Soulard.
Unknown child, in ruins at Twentieth and Papin.
Oscar Jones, roustabout, steamer Houck.
Sol Parker, roustabout, Houck.
Will Oaden, roustabout, Pittsburg.
Jacob Wendt, ferryboat clerk.

MISSING.

The list of missing people revised to date is as follows:

Arnika, Louisa, 36 years, Women's Christian Home.
Brouthers, John, 304 North Fifteenth street.
Brown, C. M., 2621 St. Louis avenue.
Buck, Mrs. Catherine, 78 years, 4368 Swan avenue.
Brandenburger, Carl, 40 years, 1827 Park avenue.
Conelly, John, Springfield, Ill.
Cooning, Alphonse, 22 years, Earlington, Ky.
Conrad, Mamie, 22 years, Jefferson Barracks.
Coles, Charles, 21 years, 4500 Washington avenue.
Cook, William, 36 years, 932 North Broadway.
Dorsey, Joseph, 23 years, 4008 Easton avenue.
Detzer, Joseph, 25 years, 1828 South Tenth street.
Davis, Mrs. Florence, 1419 Bremen avenue.
Davidson, Stephen O., 2102½ Franklin avenue.
Damat, Philip, 30 years, Wheaton, Ill.; boarded at 405 South Second street.
Decker, Jean, 56 years, 2039 Biddle street.

RESIDENCE ON COMPTON HEIGHTS.

WRECKED STEAMER.

THE GREAT CYCLONE.   235

Eastman, Fayette, colored, 1606 Carr street.
Ely, Teleph, Eighth and Washington avenue.
Forfit, Hy., 18 years, Decatur, Ill.
Fruin, Jere, 70 years, 2107 Menard street.
Foster, Sidney, 28 years, 2107½A Market street.
Gingles, Nancy, 50 years, Beaver Creek, Ill.
Goodman, Maggie, 1816 Linn street.
Goodline, L. P., disappeared from Terminal Hotel.
Gruenewald, Walter, 15 years, 1614 Arlington avenue,
Harris, Jacob, 19 years, 1114 Angelica street.
Hauck, Dr. Eugene F., 2354 Whittemore place.
Huss, Charles, 27 years, Twelfth and Cass avenue.
Heenan, Joseph, 37 years, Sixth and Poplar streets.
Hardy, Albert, 35 years, Logansport, Ind.
Johnson, roustabout, Pittsburg.
Jean, E. T., 45 years.
Jones, Samuel.
Jeremiah, Ray, 44 years; worked for E. H. Berry Boiler Company; came here recently from Belleville, Ill.
Jencks, ——, 26 years, 172 Illinois street, Indianapolis.
Keiffer, Mary, 26 years, 2110 South Seventh street.
Keim, Lizzie, 14 years, 2106 South Seventh street.
Kiefer, Charles, 1307 South Seventh street.
Kelly, E. F., 30 years, 1204 State street, Chicago; boarded at Tenth and Wright streets.
Knipp, Sadie, 507 Espenschied street.
Klein, James, 45 years, 1516 Cora place.
Lambkin, Charles Anthony, came from Cincinnati the day of the tornado.
McCarthy, Cal, 41 years, motorman.
McClellan, James, 29 years, Webster Groves.
McCarthy, Josie, 27 years, Memphis, Tenn.
Meyer, G., 60 years, Hamburg, Ill.
McLaughlin, Joseph, 26 years; worked at Ligget & Myers tobacco factory.
Moser, George, 73 years, 2830A South Ninth street.
Myers, Dora, 19 years; came from Chicago the day of the tornado.
Niesinger, A. J., Indianapolis, Ind.
O'Leary, John, 5 years, 1449 Biddle.
O'Meara, William, thought to be buried in the ruins at 1700 South Broadway.
O'Reilly, James, worked for Laclede Car Company.
Parker, M., 45 years.
Probasco, Andrew P., 816 North Twenty-third street.

Riley, Mrs. Mary, Second and Penrose.
Ridley, Mary, 22 years, Alton, Ills.
Rose, John, 19 years, 2401 South Eleventh street.
Raack, John, 50 years, Second and Soulard streets.
Richards, Ida, 11 Espenschied street.
Rutledge, Charles W., 19 years.
Reichenbacher, Henry, 37 years, 1802 Oregon avenue.
Scott, John, Seventh and Chouteau avenue.
Seitner, Raymond, 60 years, real estate agent of Peoria, Ill.
Simmons, Charles L., of St. Joseph, Mo., address any information to Simmons, Keller and Castle, St. Joseph, Mo.
Smith, James H., 25 years.
Schoff, Charles A., 1529 North Seventh street.
Schulter, Kate, 29 years.
Smith, John, 28 years, Twentieth and Penrose streets.
Sanger, Lena, 25 years.
Surgeon, James, 60 years.
Steinreide, Charles, 19 years, Cincinnati, O.
Schmidt, August, 21 years, Herman, Mo.
Struedel, F. D., 2338 South Twelfth street. This young man went cross the river just before the storm and was undoubtedly killed. He left a wife and four young children.
Thompson, S. W., 61 years, 1917 North Grand avenue.
Terrance, Mrs. M., 36 years, Nineteenth and Wash streets.
Woddell, F. H., 23 years, 6057 Horton place.
Wind, William, 40 years.
Ward, William, 27 years, Louisville, Ky.
Wetzel, Bertie, 2906½ Olive street.
Wehrfritz, Olga, 20 years.
Wagner, Mary, domestic at Planters Hotel.
Watkins, George, colored, 27 years, St. Paul, Minn.; stopping at 1608 Carr street.
Whetsel, Oscar, 4338 Fairfax avenue.
Wulfinger, John, 35 years, 2108 Bremen avenue.
Young, John, 65 years, 3007A Magazine street.
Zalijigak, Frank, Jr., 21 years, 2235 Montana avenue.
Zalijigak, Rose, 27 years, 2235 Montana avenue.

Many of the missing were certainly killed, and several were washed down the river in the flood, so that there was little chance to recover their bodies.

IDENTIFIED DEAD IN EAST ST. LOUIS.

Andres, Richard.
Anderson, Andrew
Anderson, Dick,
Brown, Joseph C., Paducah, Ky.
Bean, Mrs. Patrick and daughter.
Bruce, Mrs.
Breen, John.
Brewer, Mrs.
Barr, Mrs.
Brucker, Mike.
Bland, Robert.
Baumhoefer, August, 56 years, killed at Granite City.
Conley, Rubie,
Corrigan, May.
Clendenin, Irene.
Chapman, George, 2538 University street.
Collins, Charles.
Carroll Charles.
Connors, Mrs.
Dilligan, Mike.
Dean, Mrs.
Duffy, Joseph.
Evans, Maria.
Franks, Joseph.
Flanagan, Jim.
Free, Willie.
Fleming, H.
Frawley, Frank.
Frawley, John.
Freis, John.
Farrel, F. C.
Griffin, Thomas, ex-policeman and three children.
Gage, Mrs. Robt. and Amos.
Gladshaw, waitress Ruff's hotel.
Gladdue, Henry and Emma.
Grubb, Martin.
Hearn, John C.
Humphreys, ——.
Hayward, Scott.
Hartigan, Wm.
Heine, J. E.
Hays, John, and daughter-in-law, Mrs. Hays.
Hughes, J.
Jones, Miss Ruby.
Kildea, Michael.
Kent, John, and Ira, his son.
Kavanaugh, ——.
Keefe, M. H.
Kurtz, Jacob.
Kumre, Mrs.
Lee, Mary.
Lucky, George
Muery, F.
Morgan C.
Murray, J. W., 5234 Ridge avenue, St. Louis.
Mitchell, John.
McCann, P. J.
McCormick, Francis.
Murphy, Frank J.
McCaul, Jack, drowned.
McCormick, Francis, 32 years.
Murray, F.
Miehell, Miles.
Null, Dr. C. E.
Nicholas, P. A., Cincinnati.
O'Brien, Edward.
Porter, I. N., and son, aged 16, Brighton Ill.
Preis, F. A.
Reed, John.
Rose, Henry.
Richardson, L.
Reams, John.
Ross, Frank.
Roof, Mrs.

Rice, W. A., Huntingbird, Ky.
Rothe, Chas.
Stricker, Phil.
Snow, Sam.
Sniber, Amelia.
Surber, William.
Sage, David and Mrs.
Stock, Mrs.
Stricker, P. J.
Spellman, Martin.
Smocke, Miss.
Spricker, Henry.
Sullivan, Mrs. Emma.
Sullivan, John.
Sullivan, Mrs. James F.

Simmons, R. E.
Smith, W. F.
Steppel, Stella.
State, Mrs. George.
Trump, Dissy and Mabel, babes.
Vogt, Henry.
Valentine, John.
Valkman, Albert.
Winterman, Henry.
Wait, Charles and child.
Walmsley, Peter.
Whicher, Captain J. J.
Windley, ——.
Waldron, Mrs. E.

AFTERMATH OF THE CYCLONE.

Few people carried cyclone insurance prior to the disaster of May 27. A perusal of the earlier pages of this book will show the reason why. Within a week of that day millions of dollars of this class of risk had been written, and from all appearances the work had only just commenced. More tornado insurance was written on Decoration Day, a legal holiday, than in the entire year of 1895 in the whole State of Missouri. The last official report of James R. Waddill, Superintendent of the State Insurance Department, shows that last year's tornado risks in the whole State footed up only $1,781,589. The premiums amounted to $20,873, and the losses paid to $12,511. It was a very profitable year's business for the tornado underwriters, but the cyclone in St. Louis ate up

DESOLATED HOMES.

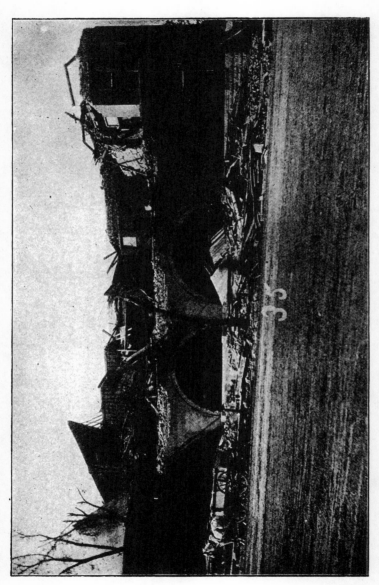

UNION CLUB.

the profits of 1895 and all previous years. The local agency that suffered most was that of Delafield & Snow. It had $45,000 on one building, Central Elevator B. The structure was damaged over $50,000, the portion above the bin walls being wrecked. The machinery on the upper floors will have to be replaced. Luckily for the United Elevator Company, there was no grain in the elevator at the time of the storm. The company owned ten elevators, but the only tornado insurance it carried was on Central B. Nine of the ten were damaged by the storm, the Venice elevator being the exception. A part of the roof of the main house of the East St. Louis elevator was blown off, the engine house and boiler house wrecked, and an addition to the elevator partly destroyed.

The grain was transferred to the Merchants' elevator. The Valley elevator was practically destroyed. Central B., as already stated, was a total wreck above the bin walls, including the machinery in the upper floors, and the river house was completely wrecked. The roof and smokestack of Central A. were blown down. The roof and smokestack and a corner of the building of the Union Depot elevator are blown away. A part of the roof and the smokestack of the St. Louis elevator were blown away, and the warehouse is almost a total wreck. The Union elevator had its roof blown off, and corner blown off the cupola and the river house and conveyor badly damaged. All above the bin walls of the Advance elevator was a wreck, and the machinery on the upper floors is destroyed and the engine and boiler house wrecked. The upper part of the river house of the Merchants' elevator is blown away.

INSURANCE COMPANY'S LOSSES.

The $45,000 insurance on Central B was divided about equally among the Royal, of Liverpool; Aetna, National and Hartford, of Hartford; and Home, Niagara and Glens Falls, of New York. St. Louis has four other elevators—Burlington, Farmers', Rogers' and Terminal. Only the Terminal was damaged, and that one not seriously. The loss is covered by tornado insurance. The Rogers' has tornado insurance, but the others have no such protection as yet.

The insurance company that suffered the most from the storm was the Concordia, of Milwaukee. It was reported in local insurance circles that the storm would cost the Concordia fully $50,000, but according to Frank Tombridge, the company's general agent here, it will not exceed $30,000. Mr. Tombridge received a telegram from the president of the Concordia to pay all claims at once—that of Mr. Eschrick, for damages to his store on Grand and Gravois avenue—and settled ten others. The policies on which the $30,000 will be paid foot up $250,-000. From February 1 up to the storm Mr. Tombridge wrote $206,000, and in the three days following it over $400,000.

He relates several interesting incidents connected with his recent tornado business. Ten days ago Mr. Ottenad, a merchant in Frenchtown, took out a tornado policy on his store. Wednesday the building was wrecked, and

Mr. Ottenad killed. His family will receive $1,700 in tornado insurance. Another Frenchtown merchant took out policies amounting to $14,000 on the same day as Mr. Ottenad. When a collector called on him last week with a bill for the premium, the merchant returned the policies with the remark that $37 was too much to throw away. His store was badly wrecked by the storm, and he wishes now that he had "thrown away" the $37.

The only St. Louis insurance company that writes tornado insurance is the American Central, the charters of the others not authorizing a tornado business. "We have 185 claims for losses from the storm," said Secretary John H. Adams yesterday. "They amount to not over $10,-000. George Ittner took out a $3,000 tornado policy on his residence on Ann avenue Wednesday afternoon at 3 o'clock without paying the premium. In less than three hours his house was wrecked. Early next morning he came in and paid the premium, and reported his loss; the first we knew of it. We paid him $1,800 before night. Our next largest policy was for $2,500, A. A. Selkirk, the auctioneer, having divided his $5,000 tornado insurance between us and the Northwestern National, of Milwaukee. The building insured was wrecked by the storm."

Mr. Adams wrote policies to the amount of $600,000 in two days. A very large number of insurers were for loan, trust and building companies. It has heretofore been their custom to insure the buildings on which they have made loans against loss by fire only. Now they are taking out tornado insurance.

Geo. D. Capen & Co. placed $300,000 of tornado insurance since the storm with the Home and Phœnix, of New York, and the Springfield, of Massachusetts. "We were up late last night writing policies," said one of the managers two days after the tornado. "We used up all the blank policies in the office, and had to telegraph for more. Before the storm our tornado business was almost exclusively on residences. The new business is largely on business houses and their contents. A good many stables are also being insured, and considerable household furniture. People laughed a week ago when we asked them to take out tornado insurance. Now they are rushing to get it. We were treated very nicely by the storm. Our policies are mostly in the West End, where very little damage was done. Our losses will not foot up more than $1,500.

INDIRECT LOSSES.

Because the tornado ruined many mills and factories, a large number of men will be out of employment for weeks and months. Children and mothers will want because their providers are idle, and strong, willing men will weep because they cannot feed hungry mouths. How many men have been deprived of their opportunity to earn a livelihood cannot be stated with any degree of accuracy, but the majority of them are skilled workmen who have made good wages and had accustomed themselves and their families to some of the luxuries of life.

True, there is a silver lining to this cloud. Hundreds of laborers are being employed to clear away the debris

ALBANY DANCING ACADEMY.

ON THIRD STREET.

and artisans are in great demand for building purposes. But this is a readjustment of the labor conditions which will prove disastrous to the mechanics and craftsmen who toiled in the mills and factories. They cannot become builders, and if they are hired as laborers they will still be sufferers, dependent upon an humble and precarious employment.

The officials of many concerns which were leveled, frankly admitted that the disaster threw out of employment bodies of men numbering from 25 to 200. Others pointed to the squads of laborers, and said they had more men on their pay rolls than they had before the storm.

This comparison is deceptive. It loses sight of the working capital that has been blown out of the business of the concerns affected. This loss of capital will affect the workmen who were employed in those mills and factories, as when business is resumed it may be on a smaller scale and with reduced forces. The weekly reports of the mercantile agencies showed how the credit men of the business world looked at the storm. Those reports contained long lists of factories with tornado damages attached.

The reports did not say that these figures of diminished credit represented losses to workmen, but that goes without saying. Perhaps many concerns are cleaning out their ruined works with money borrowed to pay the laborers. Certainly not a few of them will be obliged to ask extensions of current loans and to make new ones in order to resume business.

Time alone will tell how many concerns will go to the

wall, for the interested persons are keeping a stiff upper lip in the hope that they can withstand the mercantile cyclone that is forming, and to the credit of wholesale houses, many of them are offering extensions without being asked. One cigar store owner notified his supply house he was as poor as the day that he was born. He received by messenger a receipt in full for the amount of his indebtedness, with a notification that as soon as he could open up again in any way, his store would be stocked for him on any terms of credit he desired.

"All was lost save honor,"—no, not quite. The reputation for honesty which the sufferer had earned for himself, was sufficient capital for him to commence business on again.

SUFFERING MECHANICS.

The number of men thrown out of work cannot be estimated, but it certainly runs far into the hundreds in excess of the number employed to clear away the rubbish and repair the damage.

The largest number of men now idle on account of the damage done one concern is the force of 250 formerly employed by the St. Louis Refrigerator and Wooden Gutter Co. The factory of this concern, at Main street and Park avenue, was burned, and is a heap of ruins. It will be a long time before operations can be resumed, and as yet little has been done to recover from the blow as the adjustment of the fire insurance is proving a tedious task. Eventually business will be done on the same scale as of yore, for this company is rated very high by the mercantile agencies, and is believed to be the largest and strongest concern of the kind in the country.

Its recent order for 100,000,000 feet of lumber is talked of by kindred concerns as the largest contract of the sort ever made. The 250 persons employed in the factory of the St. Louis Refrigerator and Wooden Gutter Co. were engaged in making everything "from a needle to an anchor."

The St. Louis Steam Forge and Iron Works, better known as McDonald's Forge, at Main and Miller streets, was badly damaged by the storm. Up to last Wednesday evening it gave work to seventy men. Now ten men are engaged in cleaning and repairing. The company expects to be in shape in about three weeks or a month. Until then sixty men will be in idleness.

When a reporter called at the office of the Union Iron and Foundry Co., Second and Barry streets, groups of workmen were standing on the corners. They had visited the shop to get their pay and their tools. Similar scenes are transpiring daily about many like places.

The men got the wages which they earned prior to the storm and they looked at the money with a knowledge that it may be the last they will receive for some time.

The Union Iron Foundry Co. had 100 men in its shops. They have been idle since.

TELEGRAPH COMPANIES OVERWORKED.

The storm shut off all telegraphic communication between St. Louis and the outside world for hours, and hence the reports published in other cities of the extent of the calamity were wildly exaggerated. It was the greatest disaster of the time, but St. Louis, being a city of 600,000 inhabitants and having over sixty square miles of territory, could not by any possibility be "blotted off the map" by a tornado. Yet such was freely announced, and one imaginative writer placed the loss of life at 100,000. Telegrams and cablegrams of inquiry came in consequence from all parts.

Aside from the actual storm damage sustained by the telegraph companies they were in an absolutely paralyzed condition up to the following Sunday.

Even had not a single wire of the telegraph companies been destroyed by the tornado, they could not have handled the immense volume of business that was thrust upon them.

Even had their armies of operators been multiplied by ten, every instrument clicking constantly, and the army of messengers been multiplied by 1,000, and the whole force worked night and day, the amount of business created by the storm could not have been promptly handled.

Within an hour after the tornado brief bulletins were sent over the one or two wires remaining in working order to the outside cities announcing that a terrible storm had struck St. Louis. At that time it was impossible to particularize, for the extent of the damage was unknown.

These brief bulletins were repeated all over the United States and cabled to all parts of the civilized world. The afternoon papers of New York, Chicago, Cincinnati, San Francico, New Orleans, Boston and other large cities issued extra editions. They had no definite information at

ST. JOHN OF NEPOMUK CHURCH.

WRECK OF AN ELEVATOR.

hand, and they relied on the imagination of their writers.

These publications with their exaggerated reports were sold on the streets long after midnight, with the result that all St. Louisans abroad, and those residents of other cities in this country and foreign lands, who had relatives or friends in this city, rushed to the nearest telegraph office and started messages of inquiry. Hundreds of thousands of St. Louisans in this city, whose first thought, after their own safety was assured, was to reassure friends and relatives abroad of their escape, hastened to the telegraph offices and left from one to fifty messages to be sent, subject to the delay which they knew would be experienced.

No city in the world, in proportion to its population, has better telegraphic facilities than St. Louis. But the storm had blown down the wires and stilled the instruments, and for a few hours St. Louis was virtually in a world to itself. Thousands on top of thousands of messages were stacked up here to be sent, and as great a number were filed in other cities for St. Louis delivery.

In addition to the private telegrams inquiring about friends and in reply to queries many special newspaper correspondents had congregated in the city, and to a great extent they monopolized the wires with specials to their papers. Friday night nearly 1,000,000 words of special telegrams were sent by these correspondents.

In these hundreds of thousands of private messages, business was almost a tabooed subject. One operator kept account for a given time of messages received by him. Of the seventy telegrams he handled, sixty-eight were inquiries from relatives and friends. The other two were of a business nature.

At the Post-office the conditions were almost as bad as with the telegraph companies.

When the greatly exaggerated reports were bulletined over the country, thousands of telegrams poured into the city. They came in such a quantity that the telegraph companies were powerless to deliver them, and all day Thursday and Friday the telegraph companies put 2-cent stamps on the envelopes and dumped them by the bushel into the post office for delivery. The wires were working so badly that nearly all the replies were sent by mail, and this increased the bulk of the matter fully 100 per cent. while the force was diminished nearly 25 per cent.

The telegrams were given the preference and were sent out as rapidly as possible. While the business mail increased greatly the amount of miscellaneous letters simply flooded the office and the carriers were all overloaded.

The greatest increase in bulk was in the newspapers sent through the mails. From the newspaper offices the increase was only about 15 or 20 per cent. But the people all over the city felt themselves powerless to describe the storm in letters and they sent thousands of extra papers to their friends all over the world. The letter boxes all over the city were piled high with these papers. The volume of this kind of matter jumped up from an average of 600 pounds per day to 10,000 and 12,000 pounds.

TWO DAYS AFTER THE CALAMITY.

Two days after the cyclone, save where the city has cleaned the streets, little had yet been done by the tornado sufferers to clear up the debris. They seem still stunned by the disaster and apparently knew not how to go to work.

Thousands of visitors crowded the thoroughfares and vied with each other in securing some memento of the storm. Their industry would lead a casual observer to believe that work at reconstruction had actually begun. Such is not the case. Labor is scarce, and the house owners do not seem to know how to proceed. Many of the residences along Mississippi avenue, Jefferson avenue, Park avenue and Ohio avenue can yet be saved, but the inhabitants do not feel assured of this, and they are waiting to hear the reports of the building inspectors.

Teams of all descriptions crowded the narrow avenues through the ruins and scattered lime over the sightseers. On the green sward of Lafayette Park a company of militiamen bivouaced and sought the insufficient shade of the broken branches where once were lordly forest trees.

A few children played happily on the sward as though their dwellings were not half destroyed. The sufferers themselves gathered in knots at the doorways of the houses yet standing and spoke in quiet voices, looking passively resigned.

On the side streets every form of vehicle was in use. Furniture vans and those who could not afford to pay the high prices charged for teams were carrying the personal effects which survived the storm to places of safety.

They were all too busy to mourn longer, and a more cheerful feeling prevailed. A "St. Louis Star" reporter passed several women who were singing as they stood guard over the goods which the male members of the family were rescuing from the debris of fallen walls.

The linemen worked industriously day and night cutting down and coiling the broken wires, which hung everywhere, so that the streets were no longer snares for the unwary. Out Jefferson avenue each householder seemed to have constituted himself a member of the street cleaning brigade, and order rapidly succeeded the chaotic state which has prevailed after the cyclone. At half a dozen houses workmen were engaged tearing down unstable walls so that the work of reconstruction could be resumed. In front of a store a parrot in a battered cage, his feathers disheveled and possessing a general appearance of a hard luck tale, was noisily scolding the world for going awry. His indignant protestations over his ill-treatment caused a smile to illuminate many a careworn countenance. At 2310 Hickory street a goat, which somehow survived the storm, was trying to butt away the bricks which had covered its little domicile in Mr. John P. Ewart's back yard.

"I've had dinner and I guess the wife will have a bit for supper, so what's the use worrying," remarked Henry Oldhaus cheerily, as he scraped the mortar off the bricks of his fallen house on Rutger street, so as to prepare for rebuilding.

Where there were no injuries sustained the working people in the side streets take the calamity much more

FREAK OF THE STORM ON COMPTON HEIGHTS.

SCENE ON JEFFERSON AVE.

258

philosophically than in the more aristocratic neighborhoods and are earnestly going to work to repair the damage. For one thing, they have not suffered the loss of the thousands of pretty and valuable bric-a-brac, which represent years of collection on the part of the wealthier housewives, and then they are more habituated to hard knocks. So that it is probable that such streets as Hickory, Rutger, Caroline and St. Vincent will be rebuilt as they were before, except for lack of shade trees, far sooner than the avenues skirting on Lafayette Park.

VISITING THE CYCLONE DISTRICT.

Multitudes were attracted to St. Louis to see for themselves what the storm had done. The greatest crowds of all were on the Sunday following the calamity, when they were dangerous in their magnitude. Over 140,000 people came in by train on that day.

The visitors swarmed the sidewalks in the business portion of the city, and together with the crowd of sightseers of home folks choked up the thoroughfares leading the path of devastation in South St. Louis. Exclamations of wonder, horror and amazement were heard every where about the scene of devastation. The steady uncomfortable drizzle which set up about 3 o'clock Sunday afternoon did not serve to drive the visitors from the scene and not until nearly train time did they consent to leave the scenes of horror.

The visitors began to arrive in the city on special excursion trains Sunday morning at 7 o'clock. From that hour on until 2 o'clock in the afternoon regular and special trains continued to arrive at short intervals. Each train was loaded to its fullest capacity and in short order the midway, lower and upper sitting rooms of the station became congested. Never before in the history of the Union Station had such a large crowd assembled there except on the night of its dedication. For the first time since the opening night the upper sitting room was crowded to overflowing. In the lower sitting room there was barely breathing space, while out on the midway there was not even comfortable standing room.

Viewed from an elevation the midway seemed a moving, surging sea of hats. For a while in the afternoon the crowd thinned out, but about 4 o'clock the dense wave of humanity surged back into the station again, overflowing and flooding the entire building. The dense crowd hampered persons who wished to go from one part of the station to the other.

Despite the congested condition of the station when everybody was homeward bound, the people continued to pour into the place in a steady stream through every entrance. It seemed that the station would not hold another person, and yet thousands continued to arrive and lose themselves in the throng.

One pugnacious excursionist who objected to having his corns stepped on by anybody, created some little excitement on the midway by striking a stylishly dressed young man, with the exclamation, "Get off my corn, you dude." The well-dressed young man made an effort to appease the citizen and forced his way to other and safer quarters.

The splendid facilities afforded by the Union Station

for handling large crowds were severely tested. Superintendent Dunlop was in evidence here, there and everywhere, giving instructions to his regular force of attendants and extra corps of assistants. Station Master Jere J. Coakley was one of the busiest men about the station. Upon him devolved the duty of superintending the going and coming of passengers and the arrival and departure of trains.

The task of looking after thousands of people and directing the movements of 100 regular and forty-five special trains is an arduous one. Mr. Coakley early in the day found himself almost swamped by a host of people who could not or would not understand that an information bureau was in easy reach. It soon became necessary for the Station Master to lock his door and pull down the curtain of the window in order to attend to his regular duties. "This is the biggest crowd we have ever handled," said Mr. Coakley. "There were over 100,000 persons who came in on the excursion trains. During the day we have transferred 40,000 to and from St. Louis. At a rough estimate I would say that over 150,000 people passed through the Union Station to-day. This is a cyclone in travel."

Bernard W. Frauenthal, Master of the Bureau of Information, was probably the busiest man in the city Sunday. Questions rained in from the midway in a torrent. There was no rest. People desiring information were legion. Mr. Frauenthal was unable to make any diminution in the long line of people desiring information. He was provided with two assistants, but even then there was no getting ahead of the crowd. The more he gave answers the thicker they seemed to come. "Talk about hurricanes and one thing and another," remarked Frauenthal to a friend, "but this is hurricane, cyclone, tornado, all in one." The day was the liveliest ever spent by the persons who dispense information to the traveling public. It is estimated by Mr. Frauenthal that fully 100,000 questions were asked and answered Sunday at the Bureau of Information.

ANOTHER DISASTER NARROWLY AVERTED.

There was a time at Union Station when, for a few moments, it looked like the city was to be visited by another calamity, indirectly brought about by the more terrible catastrophe of a few nights ago. A jelly-like, squirming mass of wornout humanity was packed and jammed in the Midway passage. And that passage with its immense area of 36,000 square feet, was practically impassable. Women could be seen sitting on top of United States mail boxes, to which places of safety they had been raised by relatives or friends. Families were forced apart; women were crying. There was no room inside the building. The spacious waiting-rooms downstairs and up were but a repetition of what was taking place along the gates affording entrance to the trains. Woman fainted. There was a danger of children being crushed to death.

There were fully 30,000 people in that inclosure 50 feet wide and 606 feet in length. Something had to be done to save lives. The jam became unmanageable.

STEAMER EXPORTER.

A DREARY WASTE.

Thousands in the multitude had stood up during the night before or rode on the platforms of coaches in order to get to St. Louis and satisfy a craving that had gone forth throughout the adjoining States to witness the ruin and woe that had been left in the wake of the Storm King's trail. The constantly increasing gathering kept on crowding those already in the midway opening closer to the big iron railing, and everyone seemed to realize that at that moment it required but a very slight provocation to create a panic that would be fraught with great fatality. Men climbed up the iron railing, women shoved babies through the larger holes in the bottom of the fence to save them from the apparently unavoidable disaster. Regardless of rules of order, all the gates were at once thrown open, and the station turned over to the multitude.

Vice-President Bryan, of the Terminal Association, at once telephoned police head-quarters that the crowd had gotten beyond their control, and asked that 50 patrolmen be sent to the station as soon as possible in order that disaster may be averted. In the meantime Vice-President Ramsey, of the Wabash, had taken care of the babies and little children that had been passed through the openings in the bottom of the fence, and the several women who had fainted were removed to places of safety. Although trains, crowded to the guards, had been leaving Union Station since three o'clock in the afternoon, it seemed the jam was an inexhaustible one.

ONE HUNDRED AND FORTY THOUSAND, ACTUAL COUNT.

By actual count there were 140,000 passengers who arrived at and passed out of the gates at the Union Station on Sunday. Out of this number, however, there were, by a conservative estimate, about 100,000 passengers who came from other points and from other States to go over the wreck and ruin in this city. The other 40,000 passengers were those who returned from East St. Louis on the trains of the Terminal Association. It was expected that an immense throng of people would gather in the city, and the Terminal people arranged trains ten minutes apart between St. Louis and East St. Louis for the benefit of the excursionists who wished to visit both sides of the river. Out of twenty-three roads running into this city, five companies alone brought in by actual figures 25,000 people. The Wabash had thirteen trains in, with forty-six cars, and brought in over 7,000 excursionists. The Baltimore and Ohio Southwestern had seven trains in with fifty-six cars, and had about 4,500 passengers. Vandalia brought in 5,000 passengers on eleven trains, with sixty-eight cars. The Burlington made no specially reduced rates for the occasion, yet they had eight trains into the city, with thirty-eight cars. The Clover Leaf had only three trains in, with twenty-six cars, but they carried about 3,000 people.

The Louisville Air Line had four trains, with twenty cars. They were so short of passenger equipment that they had to bring freight cabooses into service in order to try and handle the people. The Missouri Pacific and Iron Mountain system brought in twelve trains in, with seventy-five cars. The Bluff line had only two trains in, with thirteen cars. They were hard up for cars and could not begin to handle the people. They even had to stop the sale of tickets out on the line, because they could not furnish the necessary cars to haul the passengers. On one of their trains in Union Station there were passengers actually riding on the roof of the rear car. The St. Louis and San Francisco had five trains in, with twenty-eight cars, and C., P. & St. L. brought in two trains, with twenty-one cars. The I. C. brought in five trains, with thirty-four cars. The Louisville and Nashville brought in four trains and twenty-eight cars. They had to cancel one of the excursions from points on their line because the necessary coaches could not be secured in time to take care of the passengers. The Mobile and Ohio had two trains and fifteen cars. In addition to all regular trains, upon each of which there was more or less excursion travel, there were twenty-nine special excursion trains into the city. Vice President Bryan, of the Terminal Association said there had never been anything like it before.

HALF A MILLION SPECTATORS.

The throng from outside was re-inforced by greater throngs from within the city limits, and it is believed at least 500,000 people passed along the desolated streets during Sunday. The strangers mingled with the residents on a sight-seeing tour in the storm's wake, and all day long the mighty throng surged through the streets.

The People's Railroad, which runs through a great part of the damaged section of the city and around two sides of Lafayette Park, and the Compton Hill division of the Lindell, which skirts the park on the north, carried thousands and thousands of passengers. The Jefferson avenue line, which was in operation as far south as the bridge across the railroad tracks, carried them by the thousands, while others came on foot, in buggies, wagons and in cabs.

From both ends of the city the Broadway cable hauled people by the 200 and 300 to the train, with three-car trains running in either direction every minute in the day from early morning till long after dark.

Around the principle ruins where there was loss of life or particularly heavy property damage the curious gathered till the streets were blocked, and it was all the police could do to keep the crowds moving. From every street and avenue they poured into the streets where the storm's fury was greatest, and gazed in open-mouthed amazement at the horrid picture presented.

From the West End residence portion of the town they came by the Lindell on all of its branches, the Suburban and the Olive street cable, invading thence the section laid waste through every street affording ingress. The Laclede avenue and People's lines brought their pro rata from the west and southwest, and the Cass avenue, Northern Central, Mound City, Bellefontaine, Union, Fair Grounds lines and Citizens' lines fairly emptied the populace of the northwest part of the city into the Lafayette Park and Broadway districts.

The crowd double discounted any that has ever been seen at the Veiled Prophet's parade, the Fair or the illum-

MUSIC STAND, LAFAYETTE PARK, BEFORE THE STORM.

MUSIC STAND, LAFAYETTE PARK, AFTER THE STORM.

inations, and the estimate of its number was given at from 400,000 to 500,000 by conservative men.

FEEDING THE HUNGRY.

A half dozen wagons were busy all day distributing food to the hungry. Over a hundred teams were kept busy moving furniture from the wrecked buildings to dry and comfortable quarters for those who did not have the means to pay for a van. There were hundreds of them. The most activity in that direction was in the Soulard district. Headquarters are at the Soulard Street Police Station. The Captain has given up his room to the Relief Committee, and the building was thronged all day with applicants for aid.

A perfect system had been established for the extension of aid. All applications were received and recorded by the clerks in charge. The chairman then sent a responsible person with the applicant to investigate the merits of the case. If the visitor reported that the applicant was deserving, an order was given for the needed aid. If it was a moving wagon, a teamster was given an order to move the applicant at once, and it was done forthwith. If the party needed funds to pay rent and was found worthy an order bearing the signature of the chairman was given him agreeing to pay one month's rent. These orders were redeemed with cash when presented for payment to the treasurer of the fund.

AFTER THE CROWDS HAD GONE.

Sunday night, amidst the ruins, all was darkness. Quietness reigned supreme. An occasional flash of electricity as a trolley slipped from the wire, would momentarily light up the scene. Here and there people flitted, carrying lanterns and looking like straggling lightning bugs in June. A sudden rap of a policeman's club on the sidewalks startled the passersby. Occasionally a gong on a cable car would break the stillness of the night. Sheet lightning at intervals lighted up the eastern sky and brought out into relief the frowning ruins of homes, stores, churches, buildings and clubs. The twisted stumps and broken trees in Lafayette Park made a ghostly sight. Ruins, ruins everywhere. There were no lights streaming from the windows of the mansions on the avenues. Occasionally a straggling beam would flicker out from a house not totally wrecked. All was as silent as if the city was deserted. Such was the condition of affairs last night out in the ruined residence district.

There were but few people on the streets. Policemen were everywhere keeping guard against the crooked gentry. Out Jefferson avenue could be seen the flickering lights of torches moving about as the men worked on the ruins of the power house at Geyer avenue. This was the only sign of life out there.

West of Jefferson avenue in the Compton Hill district it was as silent as a graveyard. Once the tinkling music of a piano broke the stillness for a moment and then there was silence again. The park was deserted save for the officers at the police station. Coming on down Lafayette avenue the only speck of light was at the Phœnix Brewery, where Engine Company No. 7 has its temporary quarters. The old City Hospital was deserted. Not a sign of life was visible. Occasionally a brick or piece of debris would fall, and when they struck the ruins below the crash would re-echo for several moments. Here and there lights could be seen through the broken trees in the houses untouched by the storm.

On down and the wrecked Soulard Market loomed up in all its ghastliness. One or two men stood about swapping personal experiences and telling of the heroic deeds they performed. The ruins of the old Insane Asylum frowned ominously from the north. To the south light was streaming from the Soulard Police Station. From upstairs the echo of a hymn floated out into the stilly night. 'Twas a queer state of affairs. Downstairs the police were guarding the offenders against public peace and morals. Upstairs the members of the Soulard Station Mission were worshiping their Deity. As the old hymn "Nearer, My God, to Thee," floated out on the night air a sergeant came to the door to answer a knock. He listened a moment, took off his hat until the last note of the hymn had died away, then, with a sigh, replaced it, saying: "Misery makes strange bedfellows."

ALONG BROADWAY.

Along lower Broadway there was not much life. The saloons were almost deserted. No drunken people could be seen. The calamity had sobered everyone, and most of them have remained so. On the fronts of the stores were large banners announcing great sales of damaged stock. Around the ruins of Ottenad's building a crowd of idlers gossipped of the happenings of the day.

On down Soulard street toward the river the quietude of death reigned. Beneath the ruins, there were many ghastly corpses. Over on Rutger street no one appeared but an officer, who silently surveyed the passerby, wondering what could induce anyone to wander about that dark and desolate place. Opposite the old French Market on Broadway on the vacant lot a small circus tent was lighted up. A band was playing a lively air. Through the tent could be seen the silhouettes of a few patrons who were willing to sit through the chilly two hours to while away the time. Across the street, silent and dark, stood the ruined church.

Down in the manufacturing and tenement district there was not a sign of life. All the wrecked buildings were deserted. The Relief Committee had moved the tenants to other quarters. Piles of broken brick, battered tin and other debris still littered the streets, and it was dangerous to walk about. The Levee was dead. Not a roustabout nor a Levee rat could be seen. At intervals a train darted along the elevated tracks. A few lights on the steamers told of their presence, while the swollen river rushed by on its headlong pitch to the gulf.

Across in stricken East St. Louis could be seen moving lights, telling of the activity over there. A train crawled cautiously over the wrecked Eads Bridge with a load of freight. A tug soon afterwards came puffing down the river, disturbing the spell.

Swinging back again into the wholesale district, the great Cupples buildings stood out in bold relief, with broken windows and twisted roofs. Numerous small buildings

HOPELESS RUINS.

A FALLEN GIANT IN THE PARK.

in ruins made a desolate scene in the stillness of the night. Chouteau avenue was deserted and black as printer's ink. Park avenue was like a lonely country lane. The only thing that broke the stillness of the night was the whirr of the machinery in the wreck of the cable power-house as it ran in the open air. St. Vincent avenue was like a ruined city—no sound, no life, no existence. It was late. The city was asleep, unconscious of the misery, the suffering, the woe of the morrow.

BURYING THE DEAD.

In all the wide, wide world there never was a sadder city than St. Louis was this sad Sunday. It was the day on which she buried her dead.

There were 200 funerals in St. Louis and her sister city on the Illinois side of the turbulent Mississippi, nearly every one of which was that of a tornado victim.

All day long these mournful corteges, wending their way silently and tearfully to some peaceful cemetery, were seen in every section of the city, and on the avenues leading to the principal burying grounds one procession followed another so closely as to make it look like one long, unbroken line, miles in length. It was no unusual sight to see two and even three funerals moving side by side on streets leading from the devastated districts, and such a vast concourse of mourners probably never gathered in a modern city to witness services so sad.

The scenes at some of the churches were painfully impressive, and in certain localities the sacred edifices seemed to have been converted into morgues. There were families of two and three and in one case of five persons buried from the same church or residence. In many cases whole families were killed, and their homes were piled into rubbish heaps in the cellars. Friends and relatives of the victims gave them decent burial from their own homes, or from undertaking parlors near the once happy homes of the deceased.

The priests in some of the churches preached half a dozen funeral sermons during the day. As fast as one body was carried from the edifice and the mourners had started to follow it to its windowless palace, another took its place on the catafalque, which was placed in front of the chancel rail on Friday to remain a fixture until Monday in anticipation of the numerous funerals that must be held. There were a dozen instances in St. Louis and East St. Louis where there were four and five corpses in the churches, residences and undertaking establishments awaiting the sacred offices.

Mothers, wives, sisters, fathers, brothers, the widow and the orphan, mingled their sobs and their tears over lifeless clay that until Wednesday's horror had lived and moved and loved. Grief and anguish and heart aches were plainly stamped on the faces of the sad-eyed groups that gathered about biers in many of God's houses. The lamentations of the bereaved and the soft, sweet music of church choirs, chanting the requiem, were rudely disturbed by the ring of the mason's trowel, the ceaseless chop, chop of the lineman's ax and the hammering of the carpenter as the work of clearing away the debris and bringing order out of chaos went steadily on, regardless of the Sabbath. It was a Sunday such as no great city ever spent before.

Singularly enough scarcely a bell was tolled. The task of conducting the funerals, pronouncing the last words and keeping the almost numberless processions moving assumed such serious proportions in some localities as to obscure its pathetic features to a certain extent, and reduce it almost to the plane of a business transaction.

The undertakers were scarcely equal to the emergency, and in not a few cases the tiny white coffins enshrouding infants, and even those of children 10 and 12 years of age, were taken to the cemeteries in cabs and carriages that were converted into hearses for the time being. It was sad to see infants at the breast and sunny-haired toddlers of 3 or 4 years—and there were many of them whose innocent lives were snuffed out in the storm—carted to chill, cold graves, some in almost deserted cemeteries in the southern section of the city, scarcely noticed by neighbors and friends in the general confusion that prevailed.

There were funerals, too, of those who have fared better in the matter of this world's goods; more pompous, more stately and more orderly, but none that were marked by greater or more sincere grief. Side by side, the funerals of the wealthy wended their way to costly mausoleums with those of less fortunate brethren and sisters destined for humble graves.

It was nightfall before the last groups of mourners left the burial grounds, and the wet sod was turned back over graves that will make May 27 a memorable day for years to come in St. Louis.

SAD FUNERAL SCENES.

Amid all the bustle there were striking reminders of the tragedy of the elements, and hearses stood on several thoroughfares.

Playmates when they were girls, clerks for the same firm and meeting death locked in each other's embrace. Misses Emma Cheney and Isa Horne were buried Saturday. No bells tolled as their storm-bruised remains were taken to the tomb. The church where they worshiped was leveled by the same awful wind which blew out their life light forever. The house where they had lived so many happy days was not to witness the payment of the last tribute paid by the quick to the dead. It also lay a wrecked mute witness of the mighty power which struck and spared not at all. Miss Cheney and Miss Horne were clerks at the Sawyer Manufacturing Company's building, No. 1819 Chouteau avenue, and when that building was literally blown to pieces, like many others, they had no time to escape. As they felt the building giving way beneath them, they rushed into each other's arms and so met death. When a rescuing party reached them, they were yet in that final embrace and as the same beam struck both they must have died together. The girls, neither was twenty years, had boarded at 1529 Mississippi avenue, but that house was wrecked, so the remains were taken to the residence of a relative of Miss Cheney, Mrs. B. T. Handley, 1432 Mississippi avenue. It was first intended to bury them in the same grave, but Wm. Horne, father of Miss Isa, preferred to take her body to Shrewsbury, and inter it in

FRONT VIEW OF CITY HOSPITAL.

ALONE IN THE RUINS.

the family lot. However, both were consigned to the grave at 11 a. m.

The funeral services over the remains of Miss Cheney were conducted by Rev. Mr. Williams, a Methodist minister, and were simple but impressive. A simple wreath of flowers rested on the bier. The coffin contained the simple inscription, "Emma Cheney, died May 27, 1896." The choir of the Jefferson Avenue M. E. Church sang some simple hymns, and Dr. Williams spoke feelingly of the fair dead girl, and of the uncertainty of life. The interment occurred later at Bellefontaine Cemetery.

There was grief in the family of little Patrick Gallagher Saturday afternoon, when they followed the white hearse, which bore his mangled remains from the home of his childhood, at 1729 Iowa avenue, to the Bellefontaine Cemetery. The child had celebrated his fifth birthday on Tuesday last, and on Wednesday was riding on a hobby horse which had been presented to him, when the tornado crushed out his life. The funeral services were held at the residence, and were conducted by Rev. Mr. Fountleroy.

There was no hearse and only four mourners who went with the poor torn remains of Johnny Helwood, colored, killed at 2113 Papin street. A plain white box resting on the knees of his mother and father with two brothers of the dead boy on either side of the casket were borne in a carriage to the African Methodist Episcopal burial ground. The father, Washington Helwood, had a large bandage across his brow. He had attempted to escape from the house when it fell, carrying his little son, John, in his arms. A heavy beam struck him a glancing blow, knocked the boy out of his arms and fractured his skull.

With the loss of the house the Helwood family was doubly bereaved. They suffered the child at first to be taken to the Morgue, but when the news of their loss reached the A. M. E. congregation, of which they were members, money was contributed to prevent the grief-stricken suffering the additional blow of seeing their child buried in the potter's field. Rev. John Mason preached the funeral service, which was held in the open air, with ruins of houses on every side. A large assemblage of both white and colored sufferers gathered around the bier and entered into the services with heartfelt sympathy.

AN EAST ST. LOUIS FUNERAL.

Among the many funerals there was none which elicited more real sympathy from the general public than that of City Tax Collector David S. Sage and his wife, who were buried in a single grave at St. Peter's Cemetery.

Not only the fact that Mr. Sage stood high in social, political and financial circles brought many to see his bier, who, through those channels, associated with him in life, but the more touching circumstances of his tragic and pathetic death. Mrs. Sage was the accomplished and pretty daughter of Attorney E. R. Davis. She was married to the handsome real estate man about five years ago, and the pair lived in a fine home on Eighth street. About

THE GREAT CYCLONE. 285

a year ago they rented their residence and a temporary home was established in the Strickler household.

On the fateful evening the couple had been entertained by St. Louis friends and returned to the city about 5 o'clock. A few moments were passed at Mr. Sage's office, and Mr. and Mrs. Sage noticing the threatening clouds, hastened their steps towards Collinsville and St. Louis avenues. As they reached the corner, an eye-witness says, they stopped for a moment and attempted to enter the front door of the grocery. The door was either locked or fastened in some way, and a break was made for the hallway. Just as they entered the door the awful crash came. The air was filled with flying missiles and the house began to fall forward. The couple were then standing in the landing leading to the first flight of steps. Mrs. Sage leaned on her husband's arm, while her other arm was high up on his breast, as if she were half-pleading for further protection. The request was granted without ceremony, and long before the building finally crashed Mr. Sage held his wife firmly in his arms, while hers were as lovingly clasped about his neck. In this position their mangled bodies were found, and as this was the position in which they were last seen alive, it is believed that when they noticed the awful effect of the storm upon the back part of the house and knew escape was impossible in any direction they awaited in terror and in affectionate embrace their doom.

The big building rocked and swayed before finally plunging into the shapeless mass for probably ten seconds. That space of time must have appeared as so many days to the pair, but they were surely motionless during the time extending from the storm's first impact until the building was razed.

A THIEVES' OPPORTUNITY.

Gentlemen possessed of thieving propensities flocked to St. Louis from all parts of the country to ply their nefarious business in the storm-swept districts. An entire gang of Cincinnati thieves arrived Sunday on an excursion train, and its members were promptly arrested by Chief Desmond's men and lodged behind the cold iron bars of the cooler at Twelfth and Clark avenue. These gentlemen were treated to a forced stay at the City Workhouse, and will not soon forget the experience of attempting to take advantage of the excitement and the disordered state of affairs in this city subsequent to the tornado. The names of the Cincinnati men were not made public by Chief Desmond. He had reasons for withholding them.

A number of the light-fingered gentry from Chicago also blew in with the storm. They enjoyed about two hours' liberty, and then kept the Cincinnati friends company in the dismal confines of the holdover. Neither the Chicago nor the Cincinnati crooks were allowed to make a single steal. They were nabbed almost before they had taken a breath of the exhilerating atmosphere common to St. Louis, and others who followed met with the same experience.

Kansas City and several other cities were also represented at the holdover. In fact, thieves from all parts of

MEMORIAL HOME.

VIEW OF SCHNAIDER'S GARDEN.

the United States were soon registered at that city institution, and from present indications they will remain longer than they expected; not free, however, to roam at will, robbing demolished residences and stores, but confined behind locked doors.

A large force of extra police and detectives were detailed at the Union Station and throughout the damaged district. Thieves were arrested without unnecessary delay and sent for safe-keeping to the holdover. Chief of Detectives Desmond went through the wind-swept localities every night, and the Board of Police Commissioners met in session daily, considering measures to protect the property of storekeepers and citizens whose buildings were damaged.

INCIDENTS OF THE CYCLONE.

NARROW ESCAPES, DEEDS OF HEROISM, RECORDS OF DISASTER AND DESOLATION.

Thrilling and miraculous escapes during the storm were numerous, but that of George A. Simons, second clerk of the lost steamer J. J. Odill, is specially worthy of note. Mr. Simons and First Clerk W. S. Woods were in the cabin when the storm struck. The boat was torn from the wharf by the fury of the storm, and when in midstream was struck by the Libbie Conger and overturned. When the crash came Mr. Woods called to his companion, "Save yourself, George, the boat's gone," and those were the last words ever uttered by the unfortunate first clerk, as at that instant the boat overturned and went to the bottom. Mr. Simons came to the surface entangled in wreckage, and drawing a long breath, dived under the water again, this time coming up clear of the wreck. He then swam to some wreckage which he recognized as part of the cabin of his boat, and in an exhausted condition pulled himself on top and floated down the stream.

About the foot of Chouteau avenue he passed a steamer which he recognized as the Harbor Boat, and in endeavoring to get on board he lost his hold on the portion of the cabin which had been his support, and narrowly escaped being run down. He then came in contact with a log, which he managed to get astride of, and continued his perilous trip down the river. Shortly after this he passed the Bald Eagle, but concluded to stay with the log rather than again risk his life trying to get on board, as he was now in a very exhausted condition and chilled by being in the water.

Further down he passed more boats, the names of which he could not see. And after two hours and a half he was rescued at Arsenal Island, four miles from where he started, by two men in a skiff, who landed him on the island more dead than alive. Mr. Simons says he was blown down stream a distance and then back again up stream, only to float down again during the next lull in the wind, and in this way consumed two and one-half hours making the trip of four miles. He lay scarcely able to move all night on the island, and was picked up next morning by the tug Susie Hazard and brought to the city. Mr. Simons has spent most of his life on the river, and is an expert swimmer, and to this he owes his life. First Clerk Woods, his companion, went down with the ill-fated boat, and his body has not been recovered. It is a certainty that five people perished on this boat, the bodies of three having already been recovered.

FATE OF A HARDIN, ILL., MERCHANT.

The finding of the body of Morris Fischer, of Hardin, Ill., added another to the long list of tornado victims. Mr. Fischer was the leading merchant of his town and a man of means. He was 60 years old, and left a widow, a son and several daughters.

He was in St. Louis for a few days before the tornado, making purchases and paying bills. He left on the steamboat Odill, which also carried $1,000 worth of goods he had bought in the city. Ever since the wreck of the boat anxious inquiries were made about Mr. Fischer, but it was nearly a week before his fate was definitely known to anyone, and not until ten days later that the remains were sent to his home. The body was found in the river fifteen miles south of St. Louis, near Columbia, Ill., by some fishermen, and was allowed to lie on the river bank a day and a night. The Coroner at Columbia, Ill., was notified, an inquest was held and the body buried. On it were found a few papers revealing Mr. Fischer's identity, also $7.50 in money. His gold watch and chain were missing, and for a substantial man of business the sum mentioned and the slim package of papers seem surprisingly small. While it is not as yet known that the remains were tampered with, the circumstances decidedly give color to that supposition.

The fact that a day and a half were allowed to elapse before the body was moved from the river bank speaks for itself.

As soon as the identity was made clear by his papers, telegrams were sent to the family of deceased and to the Krenning Glass Company of 820 and 822 North Fourth street, with whom Mr. Fischer was on close business and friendly relations.

The Krenning Glass Company at once notified the authorities at Columbia that a casket would be sent as fast as a wagon and team could get there, the railroad route being temporarily blocked. In the meantime the son of deceased, accompanied by a friend, proceeded to Columbia, had the remains disinterred and personally identified them. They then left for St. Louis to purchase a casket. Fortunately they first called on the Krenning Glass Company, who had already forwarded the casket, otherwise duplicate coffins would have been bought.

SWAM FOR HIS LIFE.

Had not Frank Curry, a small boy of 14 years been a good swimmer, he doubtless would now be at the bottom of the river with many other victims of the cyclone.

The boy was taken to the old House of the Good Shepherd with a seriously injured right arm. He told a remarkable story of his escape. He was standing on a coal

ANN AND CALIFORNIA AVENUES.
(A NUMBER OF PEOPLE WERE KILLED HERE.)

NEW 12-STORY CENTURY BUILDING AS IT WAS DURING THE STORM.

barge at the foot of Rutger street when the storm came up. Something struck him on the arm, and for a moment he was dazed, and the next thing he knew he was in the river. All around him the waves were dashing, while above the wind schrieked and roared. Everything was so dark, that he could not see the shore. The brave little lad did not lose his presence of mind, however, but suffered himself to be carried by the current until the darkness disappeared. When it did he found that he was only a short distance from the bank, and he succeeded in reaching it, but not without great difficulty. He was exhausted and fell upon the Levee, where he was picked up a short time later.

WAS DUG OUT UNHURT.

Fred. Mauchenheimer, son of the man who owned a saloon on the first floor of the tenement at Seventh and Rutger, where thirteen were killed, had a thrilling experience and a narrow escape. He said:

"I was tending bar for my father. Just before the storm struck the house it became so dark that I could not distinguish objects. I was just outside the bar when the building fell. I don't remember much about the first crash. I was struck on the back of the head by the ceiling, and fell by the side of some boxes that were piled up in the front part of the saloon, and saved from being crushed. I was unconscious for a moment, and when I came to it was a moment before I could realize what had happened or where I was. I then called for help, and they dug me out in about half an hour. My father was lying underneath me, but the blow from the ceiling had killed him. My mother was about 4 feet away and crushed to death. Ed. Schlunt was lying about 2 feet from me, but was unhurt, and was taken out when I was.

DUG HIS OWN WAY OUT.

Anton Hoover was in the second floor of the three-story carriage factory of Louis Neibling when the tornado struck it. The second and third floors crushed through into the first, and Hoover came out without a scratch and dug his own way out of the ruins. Neibling was in the first floor and ran into the back yard before the building fell. Vogel was also in the second floor and escaped as narrowly as Hoover did.

TRIED TO SAVE HER DIAMONDS.

When Dr. J. T. Pirtle's handsome and richly furnished dwelling at the corner of Ohio and Park avenues collapsed in the cyclone it buried beneath its fallen walls and roof Mrs. J. T. Pirtle, her mother, Mrs. Lucinda Newcomb, and daughter, Mrs. Lulu Smith. Mrs. Pirtle found strength enough to clear her way out of the debris. How she managed to scramble over the fallen wall and out of the ruin she is unable to state. Mrs. Lulu Smith was in the dining room when the crash came. She made a rush to her sleeping apartments for her valuables. She had just grabbed one diamond ring when the floor above came down and she felt herself hurled to the floor. She managed, however, to scramble to the top of the debris. Then the brave young lady and her mother set up a search for Mrs. Newcomb. They called to her but received no answer. In attempting to dig away the debris both ladies got their hands cut and bruised. Finally the wind ceased in severity and several men came to the assistance of the women. Mrs. Newcomb was found buried beneath plaster, planks and bricks and was unable to speak or move. She is 80 years old. The family slept in a neighbor's house that night and next day took up quarters on Chippewa street, near Pennsylvania avenue. Mrs. Pirtle, her mother and daughter are all severely injured. Mrs. Newcomb is confined to her bed. The damage to building, furniture and loss on jewelry totaled $15,000.

A PERILOUS VOYAGE.

John E. Massengale, general manager of the St. Louis and Tennessee River Packet Company, related some of his experiences while on a ten mile voyage on a partially wrecked wharfboat. The boat was at the foot of Locust street and the steamer Belle of Calhoun was tied to it. "We paid but little attention to the storm until it was almost upon us," said Mr. Massengale. 'It struck the boats from the east and forced the wharfboat out on the shore, at the same time badly damaging the office. We thought, however, that it would stay there, but in a second a harder blast struck us from the west side, throwing the boat back into the river and wrecking the office. Besides myself, in the office, was Wm. Peniston, Capt. John Keiflein, Miss Cecile Daly and the office boy. When the crash came Mr. Peniston rushed down stairs and clung to a stanchion. I started down, telling Miss Daly to wait till I could see if there was a safe place on the stairs. When I got part of the way down I saw the roof being rolled up and thrown across the deck, and knew that there was no place there for us. I started back, but by this time three walls of the office had been torn out and I found that we could go neither way. They joined me on the stairway, where we were slightly protected by a portion of the wall that remained in the narrow box-like stairway. By that time we were out in the river; the boats being bumped against other boats and barges that had been broken loose and in momentary fear that the wharfboat would be crushed and sunk.

"We passed and repassed the City of Providence several times and were propelled by the wind from one side of the river to the other. We finally landed on the Illinois side, about ten miles below. The Belle of Calhoun was with us all the way and afterward sunk beside the wharfboat. We got a skiff and came across the river. Leaving Miss Daly at a farmhouse, the men of the party set out for the city, and I reached home about 2 o'clock the next morning. Including several members of the crew of the Belle of Calhoun, there were sixteen of us on the boat, and, looking at it since, I can hardly see how any of us escaped alive."

DRIFTED TO THE BARRACKS.

On the day of the tornado Hugh Pattison, ex-clerk of the Criminal Court, A. J. Gallagher, C. and Fred Meyer, Fred Filch and Will Zepp were aboard their yacht

VIEW ACROSS FROM CITY HOSPITAL.

BRIDGE AND WRECK OF MARTELL HOUSE, EAST ST. LOUIS.

Gazelle, at the foot of Rutger street, overseeing alterations preparatory to entering the boat in the Alton races. When the storm came up all of the party, excepting Pattison, went aboard a coal barge for greater safety. Pattison watched the storm from the yacht, and saw the elevators and buildings along the river trembling. He saw a horse blown into the water, and the buggy torn to pieces. Boats in this neighborhood began to break from their moorings, and he then concluded to go on shore. As he was about to step from the yacht to a yawl the latter whipped into the air and whirled over a train of freight cars. He grasped the side of the yacht, which was lifted by the wind and carried over a coal barge out into deep water and sunk. Pattison went down with it. When he came up he grasped two small pieces of floating wood and tried to swim toward shore. As he passed the steamboat Madill, the pilothouse was blown off and landed on his head and back. Pattison was stunned, and again went under water, but he came up a moment later, and again struck out for shore, still holding on to the sticks of wood. The effort to get to land was too much for him, and he drifted down as far as Jefferson Barracks, where Captain Henry Leyhe, of the tugboat Louis Houcke, picked him up.

The tug itself was nearly sinking, and while the crew were working at the pumps, Pattison was almost thrown into the river again by a heavy gale which struck the boat. After effecting a landing at the foot of Arsenal street, Captain Leyhe rendered every assistance to Mr. Pattison, whose spine was injured by the blow from the pilothouse. Scarcely able to walk, Mr. Pattison started for his home at Jefferson avenue and Madison street, which he reached late at night. Dr. Marks was called, and pronounced the injuries serious, but not dangerous.

The men who took refuge in the coal barge suddenly found themselves drifting in midstream. Three of them secured the services of a fisherman and were brought to the Missouri shore. The others drifted with the barge to the Illinois bank, which they reached in safety, and then the barge sank to the bottom.

WHEELS IN THE AIR.

One of the strangest sights witnessed in St. Louis as the result of the tornado's depredations was the powerhouse of the People's Railway Company, at the corner of Eighteenth and Park avenue. The magnificent building was completely dismantled, and the expensive machinery of the cable road was exposed to the elements. It was a strange sight to see a big plant like that running while the building inclosing it was in ruins.

Thousands and thousands of people went to the place on cars, in carriages and on foot, attracted by the novelty of the sight. Operating, as it was, under the most adverse conditions that could be imagined, it is safe to say that never in the history of the road has it carried as many passengers as it did when it resumed running. The threatening aspect of the dull, murky clouds that hung overhead did not deter the crowds from flocking to the devastated district, and even when the rain began to pour down there was no perceptible diminution in the vast throng that crowded every square foot of the ruins.

At the corner of Eighteenth and Park avenue especially, the sight of the cable plant running almost in the street, with its stacks down and without even a roof over it, was so unusual and the evidence of the irresistible force of the elements so plainly demonstrated that the sightseers seemed dazed, and the combined efforts of a dozen policemen were necessary to prevent them from standing on the tracks where the trains were sailing around the curve at a dangerous rate of speed every half minute. Any number of people were brushed from the side of the tracks by cars moving either way, but those who took their places paid no attention to it.

SAVED BY A "DUMB WAITER."

The family of Henry Rohrbach, who keeps a saloon in the old People's Bank building at Broadway and Park avenue, escaped being killed in the storm in a peculiar way. Mrs. Rohrbach and her four children were on the second floor when the storm struck their house. A "dumb waiter" runs from the saloon to the second floor. Mrs. Rohrbach dropped her children into it and then followed them. Just as Mrs. Rohrbach jumped into the shaft the roof crashed in and the second and third floors were a mass of ruins. In the meantime Mr. Rohrbach, who was in the saloon below, pulled the children and his wife out of the shaft as fast as they came down. The first floor was not damaged, and beyond a few slight bruises and a shaking up, the four children and Mrs.

Rohrbach were not hurt. The shaft of the dumb waiter saved the lives of all five, as the second floor was completely wrecked. A large chimney fell into the room where Mrs. Rohrbach and her children were.

MYSTERIOUS MUNN FAMILY.

On the 16th of May P. T. Munn, his wife and twelve children left Lincoln, Neb., to make their home in St. Louis. Eleven days later the entire family, with the possible exception of one son, Frank Munn, were lying dead somewhere beneath the wreck left by the storm. Frank escaped. He sent a telegram back to his father's relatives in Lincoln, and then he, too, disappeared. The most diligent search failed to locate him. Where he is, or where his parents and brothers and sisters met their fate, is unknown.

The Western Union and the Postal Telegraph Companies could find no record of any message sent by Frank Munn, though they admitted that in the great quantity of business attended to in the last few days they might have neglected to register the sender of such a message.

The police were equally ignorant of any dead by the name of Munn, or of the visit of the living Frank Munn. The dead could not be among the unidentified, for no children's bodies were unclaimed.

THE CRIME OF THIEVES.

Thieves have added incendiarism to their infamous work in East St. Louis, and the militia promptly announced that they would shoot to kill any suspicious persons seen prowling about.

GEYER AND OHIO AVENUES.

LACLEDE GAS CO., 2D AND CONVENT STREETS.

THE GREAT CYCLONE. 307

Saturday night after the cyclone the residence of Wm. Mathews, of the Fire Department, on Main street, near Railroad avenue, was entered by thieves, who ransacked the building, taking all small articles of any value. They then fired the building. The Fire Department was immediately summoned by the discharge of carbines by the militia, and arrived on the scene in time to prevent the building from being damaged to any great extent.

HEROIC WORK DONE BY AMBULANCE DRIVERS.

In the hurry and worry incident to the work of rescue, the almost heroic work of the ambulance drivers was largely overlooked. These men braved the fury of the storm, defied the elements and even looked death in the face in the performance of their duties. Amidst tottering walls, live dangling wires, over heaps of brick and mountains of debris, these men guided their teams and ambulances in their search for the injured. No one thought of danger. Personal safety was forgotten for the moment. Suffering humanity needed surgical aid. Through them this aid could be rendered. No other incentive was necessary. Out into the fury they drove. Back they came with the wounded and dying and out again into the ruined path of the storm they dashed, ever alert for the signal of distress or the cry of an injured man. Truly no more noble work was performed by any body of men during the great emergency work.

Health Commissioner Starkloff was not unmindful of this; he is full of praise for the brave men who guided the ambulances about the city and brought the poor unfortunates to places of safety where surgical aid could be rendered. He was unstinted in his praise of their bravery. Not only did he commend the drivers but he declared that his staff was the best medical and surgical staff in the whole world. "Why, if I had to go through the battle of Gettysburg I would pick my own staff as the one which would do the work. I cannot say enough in praise of my men. I have not been able to personally thank them for their assistance, but I gladly take this means to extend to one and all my heartfelt thanks for their manly conduct in this great emergency. The employes of the St. Louis health department are the best men in their line in the world. That is no idle talk. They have proven it."

HORSE AND A LIFE WIRE TANGLED UP.

A man on horseback hitched his steed in front of the Flannigan building, opposite the Police Station in East St. Louis. In some manner a live wire which was dangling from the pole happened to touch the animal, which went nearly wild in an instant. He sprang backwards, breaking the halter, and then made a dash at the door of Schneider's saloon, smashing in the screen. Then turning, he shot across the road like another tornado.

In front of the station a number of the Belleville soldiers were sitting on the plank walk that had been torn up by the storm. They were expecting to go on guard mount in a few minutes, and were swapping yarns and cigarettes with great gusto. The frightened horse burst into the crowd, scattering them right and left like a flock of sheep. Some of the boys dropped their guns and dashed into the station, evidently thinking the tornado was making a return trip.

The unfortunate horse went clear across the walk and down some 10 or 12 feet into the lot that lies between the station and Kurrus' livery stable, and there he wandered around like an uneasy ghost all night long, no one coming to claim him. At last accounts he was still unclaimed. Singularly enough, he was not injured in any way by his fall. The soldiers returned, picked up their guns and went on duty, saying nothing.

TRUE BROTHERLY LOVE.

When men of means and prominence become laborers at the service of paupers, those who were reduced to poverty by the storm, charity becomes glorious.

Down in the wrecked district Superintendent Sampson of the Wells, Fargo Express Co. worked like a beaver all day after the cyclone. He had a half dozen of wagons at work and had personal charge of one of them. He did not hesitate to carry a mattress into the lowly quarters of the poor, and it seemed his especial delight to hand some old negro woman a box of provisions, perhaps more than she had ever had at one time in her life before.

Mr. Bauer, President of the Bauer Grocer Co., was a worker with the poor and could be seen rushing about between the Relief Headquarters at Soulard Street Station and the poverty-stricken homes, carrying baskets of food to the hungry.

Several prominent physicians were at work with the poor. Dr. Louis Drechsler, 2701 Blair avenue; Dr. Jacob A. Hartmann, 1400 South Broadway, and Dr. Albert B. Vogel, 4232 Westminster place, worked all of the day with the victims of the storm. Whenever a person or family was reported to the police as sick and unable to pay for medical attention these doctors would call on the afflicted, treat them and supply them free with all the medicine required.

If one wanted to see charity in all its beauty there was lots of it where the tornado swept.

HOW DID IT GET THE DIAL?

The propensity of the late storm to do queer things was strikingly exhibited at the northeast corner of Broadway and Park avenue, where a large street clock was perched on the roof of a building. The dial of this clock, possibly three feet in diameter and made of iron, was lying in the street in front of No. 16 engine house at 1337 South Broadway. It was unbroken. The strange feature of the matter is the fact that the hands of the clock remain unfastened. How the wind managed to screw the dial off is peculiar in itself, but how it could remove the dial without tearing the hands off with it is inexplicable.

One person who observed the phenomenon suggested that the hands were made of rubber.

CHIEF GROSS WAS ROLLED.

Eugene Gross, First Assistant Fire Chief, had a hair-raising experience in the tornado, although he escaped serious injury. Chief Gross was returning to the engine

HODGEN SCHOOL

HEADQUARTERS, RELIEF COMMITTEE.

house on Seventh street from a small fire in the South end. Just as he reached Seventh and Locust streets the bolt of lightning that demolished the gasometer on Fourteenth and Gratiot streets struck. Gross at once turned his horse south again and began a wild ride for the scene of the disaster, through the blinding rain. Huge scrolls of tin roof, heavy signs, barber poles and missiles of every description hustled about him. At Walnut street he turned west and had just reached Eighth street when the tornado picked up the horse, buggy and driver. The vehicle was overturned and separated from the animal, which was dashed against the curbing. The Chief himself was blown clear down Walnut street to Sixth, being rolled over and over on the street in all the mud and rain. He succeeded in securing refuge at this point and discovered that the only injury he received was the severing of a small artery in one of his fingers. The horse was pretty severely bruised.

HE SAVED A CITY.

One of those instances of silent, unconscious heroism, which makes the blood thrill and the eyes shine, was the work of Frank Horner, the Superintendent of the water works in saving the water supply of East St. Louis.

Had his presence of mind deserted him or his brave heart failed him, East St. Louis might and in all probability would have been a mass of smoldering ashes. A city split into kindling would have made a splendid feast for the fire fiend, but he was cheated of his booty by the clear head and iron nerve of Frank R. Horner, who never for an instant forgot his duty.

The water works are situated in the north end of the island, about two miles from the big bridge. When the storm broke, the power house was felled as with the blow of an ax, and the great mass of debris came piling around the engine which supplied the city with water.

Frank Horner was there at the time, and, reckless of danger, he ran to his engines to see what damage had been done. To his horror he found that the steam pipes were broken off by the falling bricks, and he knew what this meant.

East St. Louis lay two miles below him in ruins and at the mercy of the fire fiend. If he could not repair his engines the whole town would be wiped out.

Rushing into the street he fairly flew through the wires and broken telegraph poles, yelling with all his might.

"Fifteen dollars to every man who will work at the water works!"

He kept this up until he had a crowd at his heels and then he led them back to his battered and dismantled engines.

Throwing the ruins aside he got down to where the breaks were, and then he did something which risked the life of every man in the building, his own included.

If he could stop up the steam pipes all would be well, and sawing out four-inch plugs he ordered them driven into the gaping holes.

This done he crowded on ninety pounds of steam, and sent the welcome tide of the Mississippi water rushing through the mains to the city below just on the verge of a terrible conflagration.

It was a desperate thing to do, but it was justifiable, for it saved the city.

The dry white pine plugs swelled with the water and held fast, and the next morning when Fire Chief Purdy heard the story and realized what this man had done he threw his arms about his neck and said:

"By God, Frank, you've saved us all."

And he had, too, for the rain had stopped when the water came, and Benjamin Horn's great stave factoy next to the Hetzel mills was already aflame and would have gone but for the welcome water.

A JANITOR'S FATE.

When the storm destroyed the old Saxony school, on Barry street, between Seventh and Eighth, the entire household of the janitor met death beneath the falling walls.

The family consisted of John Locklein, aged 49, his sister, Matilda Rux, aged 56, and her daughter, Tiene Rux, aged 16. They occupied a little cottage adjoining the school at the rear. The heavy walls of the big school building gave way before the wind's strength. They fell on the cottage, crushing it like an eggshell, and buried the occupants beneath a mass of bricks and morter and splintered timbers. The bodies when recovered were horribly mangled.

The Lutheran Evangelical School was a landmark. In former years it was largely attended by the children of well-to-do Germans of South St. Louis. Of late years the school has been superceded by others and the old structure was negotiated for sale not long ago and was next year to be used as a school for colored pupils.

PENNED UP FOR SEVERAL HOURS IN A BASEMENT.

The family of Capt. Henry Sackman, living opposite to the Douglas school building on the Island, were among the many who had narrow escapes from death.

Realizing the danger of the storm, Mrs. Sackman sought safety in flight.

Gathering her children together she ran across the street into the basement of the Douglas School.

She had scarcely cleared the front gate before the second story of the brick house in which she lived blew down.

A moment later, when the little party were safely hidden in the basement, the heavy school building collapsed.

Ton after ton of stones and bricks came tumbling down, but the basement walls withstood the shock and Mrs. Sackman and her children were saved, though thoroughly drenched with rain.

Capt. Sackman is superintendent of the Wiggins Ferry Co. on the East Side, is President of the School Board on the Island and has been a member of the City Council for twenty years.

TEN DOLLARS FOR A NEWSPAPER.

Rev. Fr. J. J. Furlong of New Madrid, Mo., was a passenger on the Iron Mountain train coming to St. Louis the day after the tornado. He has many relations in St. Louis and his mission was to ascertain whether or not they were safe.

BAPTIST CHURCH, CORNER LAFAYETTE AND MISSISSIPPI AVENUES.

DOLMAN AND HICKORY STREETS.

THE GREAT CYCLONE. 319

"I have never seen such an excited condition as prevailed along the line of the Iron Mountain," said he. "At New Madrid we were unable to hear a word from St. Louis. Many persons who had relatives and friends in this city rushed to the telegraph office to wire them asking if they were safe. They could get no reply to repeated messages, and as a consequence they took the first train for St. Louis. We could not get a St. Louis paper anywhere. Offers of $5 and $10 were made for a paper at every station along the line at which we stopped. The train on which I came up was so crowded that both men and women were standing in the aisles.

"At a point a short distance from the city some one secured a St. Louis paper and then there was a tremendous scramble for it, as all wanted to see whether their relatives were among the dead. As a compromise they agreed that I should read aloud the account of the storm and the names of the dead and wounded. It was a trying task for me, as I shrank from reading the death list to those whose relatives might be named in it. It proved that there were several on the train who had friends in the list of dead and injured."

ALONE IN THE WORLD.

Of what a week before was a happy family there was but a weeping husband and father, who mourns over the graves of his two idolized daughters and his wife.

The storm in its fury did not spare the home of the Frieseke's, at 615 Hickory street. It crushed in the roof and beneath it was buried the whole family. At the storm's approach Estella and Edna, girls of 6 and 4 years, just beginning to open their eyes to the pleasure of childhood, huddled close to their mother, she hardly 30 and enjoying the fulness of life. The husband bolted the windows, secured the doors and extinguished the light, the latter to prevent the place catching fire should the house fall.

Above the crashing of falling houses all about them came the ruin of their own structure and they were crushed beneath it.

An hour later he was taken out with only a few bruises, but his wife and children had been crushed to death.

Frieseke was nearly crazed with grief. He visited the spot to-day where the tornado snatched from him that for which he had lived and with the thoughts that came to him he wept and wished that he, too, was dead.

SHE LOVED HER DOG.

When the sun crept over the river a few days after the storm its first rays fell on the form of a little girl of not more than 10 years, who was watching and weeping over the ruins or her home, at 916 Soulard street.

She was Lorna Clarkson, and beneath the wreckage of her home was buried the great mastiff dog that had guarded her from the time she could remember.

The attachment grew so great between the two that doubtless had the dog been spared and the girl buried he would have taken her place as chief mourner on the ruins.

Two men who were looking for work sympathized with the child and got the lifeless corpse of the dog from the wreck. The girl's father carried it away.

SHANTY BOATS ESCAPED.

Members of the houseboat colony along the river front in South St. Louis congratulate themselves on their narrow escape from the fury of the wind.

These frail crafts are protected by two high banks and have the additional protection afforded by a series of immense ice houses extending for half a mile along the river front.

None of the little houses were afloat when the storm came up. Their occupants could see portions of houses up on the bluffs being carried over their heads far out into the river, but they were below the deadly storm cloud.

A FAMILY NEARLY EXTERMINATED.

One of the saddest incidents of the big storm was the almost total wiping out of the Windhaus family of East St. Louis. The house in which the family resided was totally demolished. As a result, Lena Windhaus, the daughter, is dead, and Bernard and Mrs. Windhaus, the father and mother, are lying on cots in St. Mary's hospital on the verge of death. Both were injured to such an extent that their recovery is despaired of by the sisters at the hospital. Mrs. Windhaus sustained the severest injuries. Her chest was crushed in by the falling building and one leg crushed.

CRAZED BY GRIEF.

A pathetic incident of the awful calamity came to the notice of the police when a patrolman took to the Soulard Street police station a young woman who, after viewing the ruins of the house where her sister met death, became temporarily insane.

She was found on the piles of the wreckage on Barry street, between Seventh and Eighth, calling out to God in His goodness to bring back to her her sister.

The sight of a prettily-formed woman of not over 25 years, dressed in deep black, with clasped hands lifted up to the sky, wailing at the fate the elements had visited upon her sister and in an insane moment crying for her resurrection, drew a crowd about the place.

She tore the planks and rubbish aside as if to build again the wrecked home, but in this she was stopped by the policeman, who took her to the station.

An hour later she had regained her senses and was allowed to depart after leaving her name as Martha B. Wendell. She said she had come from Terre Haute to attend the funeral of her sister.

A PIG'S SOUVENIR.

About the only thing left standing unharmed on the Island the day succeeding the storm was the row of Lovingston flats on A street. They are wooden structures, about ten in number, one story high, and are so situated that the wind seems to have sailed on its airy way above them.

TEXAS AND ALLEN AVENUES.

WHERE CHILDREN DWELT.

THE GREAT CYCLONE.   325

Some are inclined to believe that the cyclone overlooked the row in its haste, and did not have time to go back and pick it up. Perhaps it did not consider it worth while, as everything else was gone.

Thirty-one tents were in use on the Island, and it was surprising to note the respectable appearance of several families who were occuping them.

Some of them spoke cheerfully of the situation, and expressed an indefinite kind of a hope that they would come out all right. One old woman who had seen better days was not so well satisfied. Near by was standing a small house occupied by negroes. The roof was off, but the vacancy was supplied by a tarpaulin.

"Look at them niggers!!" exclaimed the old woman. "Seems to me that they fare the best all the time."

Running loose on the island was a drove of fine Berkshire hogs. One of them carried with him a piece of tar roofing, which was blown squarely into his left shoulder. It protruded about eight inches, and though it shakes constantly with every movement of the muscles, the animal did not seem to suffer pain.

All attempts to capture the pig and remove the paper were futile.

KILLED ON HIS WAY HOME.

One of the saddest and most untimely deaths from the storm in East St. Louis was that of Harry F. Goodwin, a son of Editor J. West Goodwin, of the Sedalia " Bazoo."

Young Goodwin was in his 21st year. His mother died ten years ago, and after that time he made his home with his uncle, Capt. James S. Whicher, at De Hodiamont, near St. Louis.

A month ago Capt. Whicher and his wife moved to East St. Louis, but their stay was only temporary, as they intended to go to Indiana to live with their son.

Young Goodwin had his possessions packed, and only a few hours after the storm he would have returned to his father's home in Sedalia to live.

Capt. and Mrs. Whicher were both killed. Harry was caught in the wreck, and for five hours was pinioned down in such a position that he could move nothing but his hands and toes.

A deep cut on the back of his head proved fatal, and he died after several hours of delirium.

The body was taken in charge by his brother, Ben. J. Goodwin, a clerk in the Merchants-Laclede National Bank, St. Louis, who was with him when he died, and was shipped to Sedalia for interment.

Mark Goodwin, another brother of the deceased, is a newspaper man at Denison, Tex. He formerly worked in St. Louis.

CAPT. M. M'MAHON'S EXPERIENCE.

There is no more competent police captain in St. Louis than Martin M'Mahon, who is a humorist and a philosopher as well as a first-class organizer and disciplinarian. It was at first feared he had been killed but he was not. Indeed no man worked harder the night after the storm and for days following to help the sufferers. A few yards east of the lake in Lafayette Park stands a rustic summer house with thatched roof supported by plain pine posts. Within a few feet of this fragile structure immense trees were uprooted and their branches torn into splinters, while the summer house remained untouched,

THE SUMMER HOUSE IN LAFAYETTE PARK.
(UNHARMED BY THE STORM.)

and stood in the same condition that it was before the storm. Everything else immediately around it was wrecked and ruined. During the storm Captain M'Mahon, who is in charge of the Police Station in Lafayette Park only a short distance to the south-east of this summer house was at his post. The Captain says that when he got home after the storm he found his own house in ruins, and learned from the ladies of his family that during the height of the tornado they were all praying earnestly for protection when suddenly the roof was lifted from off their heads. While relating the incident the Captain smiled and seemed to think that perhaps the ladies prayed too hard. In other words they lifted the roof.

A HUMOROUS INCIDENT.

The disaster on the river was not without its humorous features. The ferryboat Andrew Christy, of the Wiggins Ferry Company's fleet, was one of the many crafts that drifted helplessly down the stream and lodged against the eastern bank. Afterward, the steamer Belle of Calhoun and a wharfboat floated down, and, after bumping against the Christy, passed on. Several persons were on the wharfboat and a few on the ferryboat, among them the engineer. When the boats touched, some one on the wharfboat asked the engineer if he wanted to get off his boat.

"Yes. Wait till I get my dog," confusedly answered the engineer.

He got the dog and managed to get on the wharfboat before it got by.

THREW HIM A LIFE LINE.

Harry Pence, a young millionaire from Minneapolis, had an experience during the storm, the memory of which

UNITY CHURCH.

VIEW AT 12TH AND GRATIOT STREETS.

he will long retain. Mr. Pence had been cruising about the Gulf of Mexico in his yacht Roamer. He was on his way home. He stopped at St. Louis to visit his cousin, Dr. Charles Ellis, the Dairy Inspector. Shortly before the storm he went in his yacht for a short cruise along the river bank. When the storm was at its height he was opposite Carondelet. He ordered the captain to make for shore, but it was impossible. In another instant Mr. Pence, who was on the promenade deck, was hurled in the air, and, after being tossed about by the wind, landed in the water. Some men on shore threw ropes to him, and, after some effort, landed him on terra firma. The yacht was caught a few blocks down the river.

TALE OF TWO FAMILIES.

The five persons composing the families of J. W. and J. L. Tinker, who were en route from Alabama to Wabashaw, Wis., had an experience on the river during the cyclone which they will always remember. They arrived on the steamer Mayflower, and at once secured passage on the Pittsburg, which, it will be remembered, was one of the boats that was so badly wrecked. When the tornado came down upon the river, the Pittsburg was tied up to the Diamond Jo wharfboat. The force of the wind blew the wharfboat from its moorings with the steamer far out into the current. There the steamer was exposed to the full force of the storm, and it was feared that she would go down every moment. The wharfboat was pounding her badly, and had it not been cut loose it would have torn a hole in the Pittsburg's hull.

As the steamer passed out into the river all the state rooms on the starboard side were blown away, and the Tinkers were saved only by being dragged to the port side. It was not long until the state rooms on this side fell in, and all the members of the two families were more or less injured. All that remained of the staunch craft was the hull and floor of the cabin. The texas and pilot house had entirely disappeared and have not yet been found. Mrs. J. L. Tinker, who narrated the story, says that had it not been for the captain, she and all the others would have been killed. Almost every member of the crew deserted before the storm in its fury swept down, and the captain, too, was asked to leave, but refused. The boat drifted four miles down the river and grounded on an island, and at a late hour a tugboat came to the rescue and brought the imperiled passengers to the city.

RESCUED BY AN OFFICER.

There is an anecdote of the storm in which Detective Meaney played the leading role. He was at Eighteenth street and Ohio avenue, when the storm was heaviest. The building at 1806 Ohio avenue was tottering. The occupants on the second floor, Mrs. Gorman and two children, were screaming for assistance. Detective Meaney ran up the steps of the building and got the family to one side of a wall which he figured would be the safest. The tornado took the roof off the building and the debris fell immediately in front of them. The detective then took the children in his arm and carried them downstairs. He then went after the mother and landed her safely on terra firma. All then went across the street and sought shelter in a solid building. Fifteen minutes thereafter the husband of Mrs. Gorman, who is a conductor on the Fourth street cable line, came home. His eyes moistened as he saw his home was demolished. He asked piteously of bystanders for the fate of his family. Detective Meaney went across the street to Mr. Gorman and pointed to the latter's wife and two children safe on the other side of the street. The husband wept in his joy and showered thanks and blessings upon Detective Meaney

MRS. BENE'S EXPERIENCE.

The terrible ordeal through which Mrs. Lillie Bene, of 1418 Ohio avenue, passed is perhaps without a parallel among the entire list of cyclone sufferers. Her escape from death is most remarkable, but the wonder is that the poor woman has not lost her reason. One of her children was killed before her eyes, the other buried with her in the wreck of her home and seriously injured, while she, in addition to numerous wounds, narrowly escaped being roasted alive by the burning debris.

Her husband, James Bene, is employed by the Emil Wachter Hardware Company, and when the tornado came she and her two little boys, Sylvester and Oliver, five and three years of age, respectively, were the only occupants of the residence. As the storm increased in fury she took the children and went into the middle room. Sylvester in his fright ran to the kitchen in the rear. A moment later he commenced to cry to his mother to come to him and she did so, only to see him struck by some flying debris and killed. She was too horror-stricken to think of her own safety, and as she advanced to pick up the little mangled body one of the walls fell and she and Oliver went down with the wreck to the floor below.

Strange to say, she did not lose consciousness and at once tried to extricate herself. Her hands were free, but her body, which was bruised and bleeding, was pinned fast. So high was the wreckage piled above her that all was in total darkness. She tried to call for help but could not. Her mouth and nose were filled with dust and mortar to such an extent that she was almost suffocated. With her fingers she removed the particles sufficiently to be able to breathe and then feebly cried for help. To add to the horror of her situation the timber beneath her legs commenced to burn, having caught fire from the stove of the kitchen on the ground floor. Had it not been for the mass of brick and mortar which was scattered about and which in a measure smothered the flames, she doubtless would have been burned to death before rescue came. Close to her lay the dead body of her little son and several times she managed to touch his face. Her baby, Oliver, was lying at a greater distance away, but by his cries she knew that he had escaped death. For over an hour the mother underwent this terrible ordeal before two of the neighbors heard her cries. It took them another hour to clear the timbers away. While they were at work a fireman came and dragged Oliver from the wreck. Across his body was the dead

A SAD HOME COMING.

A PICTURESQUE RUIN.

body of the little fellows pet dog—an Irish setter. The dog's body had shielded that of its master, and it was due to this fact that Oliver was not killed.

Mrs. Bene was carried to the home of a neighbor. She was moved no less than three times before her arrival at St. John's. Her body was covered with cuts and bruises and her legs from the knees down are badly burned.

A JOCKEY ON ICE.

Jockey Kuhn, a rider at the South Side race track, was in the neighborhood of the track when the storm blew up and ran to the Union Depot Railway power house for protection. He says that he saw at least twenty people running toward the building, several of whom carried bicycles. Kuhn left there and ran to a butcher shop on Jefferson avenue and crawled into an ice chest. The German proprietor made fun of him for climbing into the box and called him a coward. When things began falling around the neighborhood the German himself made a dive for the ice chest. The jockey told him to stay out, as he was too brave to hide himself. The butcher thought otherwise, however, and climbed into the chest with Kuhn.

CONTRIBUTED A $50 SUIT.

A West End resident contributed his new $50 suit of clothes to the storm sufferers against his will. A fashionable coupe drove up to the Merchants' Exchange and a footman in livery brought a large bundle of clothing to the Relief Committee headquarters. The bundle was not opened, but, with several others, was dumped into a distributing wagon.

About an hour later the coupe drove up again and the footman, greatly excited, rushed in and asked for the bundle he had brought earlier in the day. When told that the bundle had been sent out his distraction knew no bounds. He said that his employer's new $50 suit of clothes had been placed in the bundle by mistake by the mistress of the house. The suit had just come from the tailor and the box containing it had not even been opened. As no record was kept of the clothes sent out, it was impossible to recover the suit, and the footman left dejected. He declined to give the name of his employer.

HE'LL NEVER SAY DIE.

As an illustration of the determination of the average East St. Louisan to overcome the destructive work of the cyclone and an indication that the city will soon recover from the period of gloom that hung over it for a few days, the conduct of ex-Supervisor Patrick Kelly may be cited.

Mr. Kelly at the time of the cyclone and many years previous, conducted the Riverside Grocery, on the Levee. By thrift and strict attention to business he had acquired quite a competency, which he invested in the building of a row of frame buildings on the Island.

The cyclone blew down all these buildings and left Mr. Kelly almost destitute. He, however, went to work after the storm had spent its fury and erected a shanty where his grocery formerly stood.

Dr. J. L. Wiggins, an old friend of his, happened to meet Mr. Kelly at his improvised store on Tuesday morning, and while satisfied that things looked rather poorly, asked Kelly how he was getting along.

"First rate," said Kelly.

"How's that?" asked the doctor.

"Why, you see, doctor," was the luckless man's response, "I have already doubled my stock. On Monday I started in with a box of soda, and now I have a box and a half and thirteen pies. I will pull out yet."

PLUCKY GIRL.

The ruling passion of one sorrowful little girl proved strong even when the tornado had left her without a home and with only the clothing she had on. She was found by an officer hovering near her wrecked home on Seventeenth and Geyer avenue industriously turning over bricks and debris in search of her school books. She was about to graduate in the Normal School and was fearful the loss of books, which she could scarcely afford to replace, would mean failure in examinations.

A COINCIDENCE.

During the fearful tornado a window over the altar of the Mount Calvary Episcopal Church was blown down. The window was placed there in memory of James B. Eads, the great engineer who built the famous bridge which bears his name. By a coincidence, the east portion of the bridge and the memorial window fell at the same moment. The window contained a picture of the

Savior, surrounded by a crown of thorns. Underneath was the inscription, "In memory of James B. Eads. Born May 23, 1820. Died March 8, 1887." It was the only memorial window in the church that was broken, and members of the congregation are commenting upon the occurrence, the massive bridge which was the engineer's greatest work, having suffered damage in the same storm.

PINIONED IN DEBRIS.

Martin Martell, proprietor of the Martell Hotel, East St. Louis, recovered from his injuries in a few days sufficiently to visit the wreck of his house, although so badly hurt that he was reported dead. A number of traveling men met him at the Relay, and after they had congratulated him upon his escape with life he told the fearful story of his escape. He said that when the first shock came he left his dining room and came out to the front room, facing the Relay. Here he met Judge A. W. Hope, of Alton, Judge Foulk, of Vandalia, and Messrs. William Flynn and John Mumme, of Edwardsville. Hope, Mumme and Flynn decided that they would run over the roads, while Mr. Martell and Judge Foulk concluded that it would be safer in the house. The three gentlemen had scarcely left the building when the house was razed.

Martell does not know what occurred afterward to the others, but when he revived he was pinioned in such a way that his feet were elevated at an angle of probably forty-five degrees, while his head was pushed far up on

REPUBLICAN CONVENTION HALL.
(PARTLY UNROOFED BY THE STORM.)

A PROMISCUOUS WRECK.

his breast. He could not speak or see, but could hear and feel people passing over him in search for the dead at the back of the house. The fire had then started, and the horrors of a burning death confronted him. He was also further dumfounded to learn that several parties noticed him, but seeing that he was motionless, said that there was no use bothering with that dead man; let us go on and succor those who may be alive on the fire's side. In this way he believes at least fifty people passed over his body, all injuring him by tramping on the boards which bound him, and not one believing that he was alive.

At last, by a superhuman effort, he chanced to move a foot as a rescuer walked along, and that movement no doubt saved his life. The man held his lantern down closer to the form, and realizing that the boot was not shaken by the breeze, he called for help, and soon extricated the semi-conscious ex-Treasurer of the City of East St. Louis. He was taken to the hospital immediately and tenderly cared for, and is now ready to commence the rebuilding of his place at the Relay. Judge Hope was also present last evening, and listened to Mr. Martell's tale. He, too, had a trying experience at the door. He said the party had scarcely gained the outside when he was knocked down and lay upon the curbing near the railroad tracks. At that instant the house crashed in, and missiles of all kinds were piled up about him. He partly raised himself, and in that instant saw the roof of the Martell House and other large articles pass over his head and lodge near the Relay. Another instant, he was looking at a waste where a few seconds before stood fine brick and frame buildings, the Vandalia round house and other railroad property, all of which cut off a view of any part of the bridge. Now, however, he saw the structure in its entirety, and could not, on that account, realize his position. Seeing the bridge, he naturally concluded that it was he who had been blown away, and not the obstructions. Yet, turning again, he recognized the remains of the Relay Depot, and then for the first time he concluded that the storm had actually taken everything from the creek to the Levee, a distance of nearly a mile. The rain was pouring down in torrents, the thunder rolled, and the lightning momentarily lit up the heavens. During the intervals, however, the darkness was intense. These flashes set out the whole panorama of destruction more vividly than if the light was continuous, and surely, he states, the scene was more terribly impressive.

A BARBER'S STORY.

Upon the top floor of the Wainwright building there is a neat little barber shop, presided over by Louis Tisch, that has had lots of trade since the cyclone, because a great description of the tornado and its flight to the city is served up with every shave. With a haircut Tisch and his assistants point out different places of interest in the path of the storm, as they are seen from his elevated perch, and if a man takes a shave a haircut and a shampoo, maybe if he is good he can go out on the roof.

The barbers in the Wainwright building, Tisch, G. C. Adams, John B. Huppert and A. Rust, saw the storm from start to finish, and they tell a most remarkable story about it. They say it was not a funnel shaped cloud such as is commonly pictured as being the shape of a tornado. Each solemnly swears it was a horizontal black cloud that moved through the city with a twisting motion like a screw, faster than any railroad train that ever ran. Preceeding the black cloud was a dense yellow cloud that looked as though its interior was a mass of flames. From out of this cloud shot long firey arms in every direction, and wherever one of these arms struck something went to pieces.

Tisch compares the cloud to a big serpent that wriggled along up in the air and trust out a multi-forked tongue as though in anger.

Shortly before the storm broke, Huppert went up on the roof and came back with the information that there was a tornado in sight. Rust followed him and came back with a confirmation of the report, and then the two barbers went out and saw the grand marshalling of the storm in the western skies. When the rain began, they came down into the shop, and the last they saw as they were coming through the scuttle was the advance guard of the tornado as it came in from the southwest.

The barber shop is at the southeast corner of the building, and all around it are little windows, round, like the port holes in a ship. The barbers stood at the south windows and watched the tornado from the time it appeared, away off to the southwest, until a portion of it rolled up against the building and made them wish they were somewhere else. Tisch says they saw houses and business blocks go down before it, their view of the destruction it was wreaking being made plain by the yellow clowd of fire that preceded the storm proper. He is sure it crossed the river some distance below Park avenue, switched around when it got nearly to the Illinois shore and started directly up the stream. In this he is borne out by the statements of the others who were watching it.

Just as they were getting ready to move around to the east windows, in order to observe the passing of the storm up the river, a gust of wind and rain that shook the building came along, and they were in the midst of the storm. When next they saw the river and the city below, the storm had passed, and the rain was falling straight down. They saw dozens of wrecks floating down the stream, and on the other side saw all the steamboats blown away from the harbor and piled up along the bank. Then came the second storm, followed by the St. Louis Wooden Gutter Company's fire, which they saw from their airy observatory. It was late when they went down, after three hours' of uninterrupted excitement.

A colored boy named Mose is one of the valued attaches of the shop, and he was one of the spectators when the awful cloud was first seen. He instantly started for the ground, and he got there in a hurry. He forgot about the elevators and made a slide, it is averred down ten flights of stairs. No amount of persuasion could get him back to the shop that night. Next morning when he was being twitted about having run away from the storm, he remarked:

"Oh there was others."

OUTSIDE VIEW, LIGGETT & MYERS TOBACCO FACTORY.

SECOND VIEW, LIGGETT & MYERS TOBACCO FACTORY.

HER DAUGHTER'S ASHES.

Bent with age and with feeble steps an old lady hobbled into Chief Harrigan's office at the Four Courts two days after the tornado, carrying on her arm a market basket in which was a curious looking can or urn. Poverty had pinched the features of the aged woman and sorrow had left its deep lines. Secretary Espy was at his desk and he noticed the visitor and asked her what she desired. Her story was most pathetic, and the secretary was moved to pity at its recital. With hesitating words she stated her mission:

"My name is Julia Weise, and I live at 416 Barton street. Two years ago my daughter, Mrs. Louis Sehr, died in Memphis, and this urn, which I have in my basket, contains all that remains of her, after her body had been cremated. It was her last desire that her husband, Louis Sehr, should have her body cremated. For two years he carried the ashes of his dead wife with him wherever he went. Some months ago he came to this city, a sufferer from consumption. He lived with me at 416 Barton street, and did what he could to earn a living, though growing weaker every day. Finally he was compelled to remain in bed and for weeks he lingered, wasting away to a shadow. Wednesday he died, and I had not the means with which to bury him, so his body was taken to the Morgue, where it now is.

"I am greatly in need of help," said the old woman, with quivering voice, "and I want Louis buried decently. He has a brother in Utica, N. Y., Michael Sehr, a well-to-do barber, and if you could notify him of Louis' death and ask his assistance he might send money enough to pay for the funeral."

The old lady was told to come back at 4 o'clock in the afternoon. In the meantime, Chief Harrigan telegraphed to the brother in Utica, stating the circumstances, and asking if he would defray the expenses of the burial.

The urn containing the ashes of Mrs. Weise's daughter remained in the Chief's office all day. The old woman left it there for safe-keeping. She had no special place to stay, and is afraid she will lose the urn. It was the object of much curiosity on the part of attaches of the office and visitors on account of its peculiar shape. It is made of heavy tin and has a screw top, which can be removed at will. The ashes weigh about five pounds and are of a light straw color.

A MAN LOST HIS REASON.

At the old police station in East St. Louis, the officers wrestled with a man who said his name was Thomas Phillips. Phillips was picked up while aimlessly walking about the railway yards and lodged in the station. As soon as he was brought in he displayed unmistakable signs of insanity, and attempted to make a break for liberty. He contended that the officers wished to murder him on account of the part he had taken in the late tornado, and as he was a man of standing he would sell his life dearly. The next instance he quieted down and said that his name was Thomas Phillips; that he lived at Greenville, Mo., but afterwards held that he had just arrived from Canada. As soon as he was placed in the cell he fought desperately for freedom. Several policemen ran to the assistance of the turnkey and the man was finally overpowered.

He then paced up and down the corridor and denounced in unmeasured terms the action of the police in firing at him so often. The police believe that the poor fellow is a victim of the storm. Some claim that he was seen on the streets a few days ago, and if he is the same person something since has caused him to become demented. He is a fine looking fellow, tall and stout, with light hair, sandy mustache and blue eyes. His hands do not indicate that he is a laborer, and in his lucid moments he uses language with characterizes him as a man of fair education.

BATTLE OF THE WINDS.

H. D. Sexton, the wealthy real estate owner, Vice-President of the East St. Louis Electric Railway, and director in the Workingman's Bank, had a narrow escape in the storm, and gives a graphic account of "The Battle of the Winds," as he terms it.

Mr. Sexton, with his stenographer, Miss Rose Taylor, Cashier Isch of the Workingman's Bank and several others were sitting in his office on Broadway when the skirmish line of the infuriated elements hove in sight. They started to go home, but saw that they would be overtaken on the way and concluded to wait awhile.

For a time they tried to hold the front door shut, but the glass in the broad window crashed in and stampeded them. They dodged under tables and desks, and Mr.

Sexton says that above the roar of the winds he could hear some very audible and fervent prayers. A telephone pole broke just in front of the office, and was shot straight as an arrow against the wall, knocking a hole in it, and added to this the terror-stricken occupants could hear the roof crackling and falling above them.

They dared not venture out, so lay prone upon the floor, expecting the worst, but when the storm had spent its force, all rose, shook the wet mortar from their clothes and ventured out.

Mr. Sexton says that for fully ten minutes before the storm reached East St. Louis he stood in his office and watched the grand elemental display that was playing such havoc in South St. Louis.

He said at times the heavens seemed to brighten and he could see the storm moving northward at a rapid rate. He expected it to reach the East Side, but for the moment his attention was turned to a terrific gust of wind that came from the north.

The contending forces were advancing to meet each other, and they did with terrible force just about at the Eads bridge. Nothing but such combined force could have shaken and crumbled a span of the great bridge, is Mr. Sexton's opinion.

The south wind was the stronger, but the brief fight the north wind gave it was deadly and terrific. Both aimed their wanton forces upon the defenseless victims on the Levee and upon the railroad warehouses in that section. That is why the result was so calamitous. The south wind, in its maddened triumph, swept on with ter-

LIGGETT & MYERS NEW TOBACCO FACTORY.
(UNFINISHED AT TIME OF STORM.)

TRINITY CHURCH.

rible zeal and devastated every obstacle that human hand had raised to impede its march of destruction, and spared not life itself.

"It was indeed a battle of the winds," continued Mr. Sexton, "and as a common enemy, East St. Louis will wear its scars for years to come.

FAMILIES BROKEN UP.

John Bergeest, driver for the Louis Ottenad Furniture Co., at 1700 South Broadway, leaves a wife with a young boy and girl, who are suddenly cast upon the sea of adversity, with no means of support. He was by an open window trying to quiet his team when the building collapsed and his body was not found until Friday evening.

William Ottenad, president of the Ottenad Furniture Co., was in the office at the rear of the store when it collapsed and was instantly killed. His family consists of his wife and seven children, ranging in age from one week to eight years. Mrs. Ottenad is very low, indeed, and her death is momentarily looked for. The deceased left no estate, his all being in the furniture he had in stock.

Mary Rux is the only surviving member of the family which lived at 716 Barry street. There were in the building at the time her mother, Matilda Rux, 56, her sister, Jennie, 17, and Tina, 16, and a friend, John Labien, aged 49. All were sitting in the front room when the first blast came and it almost lifted the house from its foundation. Mary jumped through the door and escaped by the narrowest margin. All four of the others were killed instantly. The survivor is 19 years of age and self-supporting.

A TERRIBLE SCATTER.

John W. Dunn, Assistant City Treasurer, tells a remarkable story that he dug up while visiting the cyclone district near Lafayette Park. A driver for an ice wagon was last seen entering an alley back of the power house of the People's Railway Company, seated on his wagon with his whole outfit intact. Later he was missing, and for hours nothing was known of him. Mr. Dunn says that the entire outfit was discovered, the wagon in one place, the horse in another, and the driver in another. They were at the three angles of a triangle, and Mr. Dunn is authority for the statement that they were 100 yards apart. It is impossible to explain how the driver was blown from the seat, or the horse separated from the wagon, but the facts Mr. Dunn says, are as stated above.

A MAN AND WOMAN BLOWN AWAY.

Architect Isaac S. Taylor tells of a dramatic scene which he viewed from his office on the ninth floor of the Columbia Building, at Eighth and Locusts streets. Mr. Taylor says that soon after the storm broke, a man and woman turned into Eighth street from Locust. As soon as the rain began to fall the man put up an umbrella. When the speed of the wind increased both of them sought the railing around the Custom House, with the umbrella still up. The latter was soon wiped out by the gale, and then the man and woman were seen desperately clinging to the iron railing. The fury of the gale increased from forty to fifty, sixty, seventy miles an hour. There was a moment of darkness as the wind blew furiously, and the man and woman were still clinging to the rail. In another minute the street lit up and they had entirely disappeared. Mr. Taylor says he does not know what became of the couple, but there is no doubt they were swept down Eighth street by the wind, and it is probable they will figure eventually among the number of missing persons who have not been accounted for since the storm.

HORSE RESCUED ALIVE.

Workmen on the ruined Strickler Building came upon a live horse Sunday night after the wreck. The poor animal was freed from its predicament some time during the day.

The Strickler Building was one of the worst used by the storm of any of the structures in East St. Louis. It was absolutely razed to the ground and at least four people killed. Among those who lost their lives there were Special Tax Collector Sage and his wife, Phil Strickler, the driver, and Dr. C. E. Mill, perhaps the most prominent victims of the disaster.

There was a horse stabled in the basement. He used to draw one of Strickler's grocery wagons. After the building fell and people began to think again, some of them remembered the poor animal, but thought, of course, he must be dead, covered as he was by the great mass of debris. The workmen had been laboring ever since to get the wreck cleared away with a view to finding any more dead bodies that might be there.

Sunday night they heard sounds indicating that something was alive somewhere in the debris. Great excite-

EAST ST. LOUIS ELECTRIC CAR BLOWN FROM THE BRIDGE.

ment followed this discovery. Labored breathing could be heard, and the sound of something struggling as though to get freed from the crushing weight.

Efforts were redoubled by the men. They felt sure they were about to come upon some poor torn and bleeding man or woman, but when the last stick was lifted the dirty nose of the horse was seen.

In some miraculous way the animal seemed to be in pretty fair spirits. A bridge of timbers had formed over him in such a manner as to keep him from being mashed to death. Monday morning a quantity of oats was let down to the lucky prisoner and a bucket of water was sent after it. The poor creature was famished and ate and drank ravenously.

Great care was exercised by the men to prevent the props from tumbling out, and the horse was finally gotout unhurt.

IT'S AN ILL WIND.

The luckiest man in East St. Louis is A. M. Meintz. The tornado actually did him good instead of harm and that without any injury to his fellow-townsmen or at their expense in any way.

Meintz is a very wealthy man and also a very shrewd one. He owns a two-story building at Third street and Missouri avenue. Recently he decided to add an extra story to it and was just getting ready to remove the roof for this purpose when along came the cyclone and took it off for him.

Meintz is one of the few men in East St. Louis who had tornado insurance. His losses are entirely covered in this way and he saved the expense of getting ready for his third-story addition.

He is going right on with his plans and workmen are now busy putting on the third story.

TYPICAL SADNESS.

One of the most pitiable cases seen at relief headquarters was that of Mrs. August Brauer and five of her six children.

Until the storm they lived at 500 Broadway. Their house was torn down completely. Mr. Brauer was so badly bruised that he could not walk, and Philip, 8 years old, had a leg broken.

How the other members of the household escaped is a marvel.

Not a single article of furniture was saved and all of their clothing was lost.

To add to their misfortune the baby was badly bruised on the head and has since become sick because of the weather.

The family found temporary shelter with Mr. Brauer's sister, Mrs. Andrew Peterson, whose husband is a member of the East St. Louis Fire Department.

A WILD AND AWFUL RIDE.

Two maddened horses racing south on Jefferson avenue, dragging behind them a "trouble wagon" of the Union Depot Street Railway Company, with two frightened men clinging to the rails on the elevated tower which rests on the bed of the wagon, was one of the scenes witnessed.

The wagon had gone to 2313 South Jefferson avenue to repair a broken trolley wire. The wagon had driven into the tracks. The tower was elevated so that the two men on top could be in easy reach of the trolley wires.

After the wagon was in place and the two men were at work on the wires the driver left his seat to go to the sidewalk. He had hardly reached the ground when the wind blew a paper in front of the horses. They became frightened and started on a dead run down the avenue. One of the men in the tower had his hand in the ropes about the trolley wire and his arm was nearly jerked off.

Both men screamed, but that only seemed to urge the horses to greater speed.

So high was the tower that the rails nearly caught in the overhead wires, and a dozen times the endangered men stooped and crawled to the floor of the high platform to keep their heads from being severed from their bodies.

Men who were blocks down the street saw the team coming and realized the jeopardy in which the men's lives were put on the tower and they rushed frantically into the car tracks, some pulled their coats from their backs and waved them in the air in an effort to stop the horses, but that only made them turn from the tracks and the wheels, in gliding against the rails, nearly overturned the top heavy vehicle.

Finally one of the two men who had been dodging electric wires and boughs of trees scaled the rail and while the horses were going the fastest lowered himself down the tower until his feet touched the bed of the wagon. Then he sprang to the ground and rolled over into the gutter.

The horses made for the car sheds and dashed into one of the car doors. The tower with the man on it struck the top of the door and it was torn loose from the bed and toppled over to the sidewalk. The man jumped and struck in the street, sustaining nothing more than a sprain of the ankle.

THE WIND'S IDIOSYNCRACY.

The twister fairly out-twisted itself when it demolished Mt. Calvary Church at Jefferson and Park avenues. The

SCENE ON CHOUTEAU AVENUE.

belfry was on the northeast corner of the building before the storm. Afterwards it wasn't. The tornado struck the church and carried the roof up into the air. While it was up there the walls gave way. Then the roof turned end for end and dropped on the ruins, covering them completely. When the storm had passed it was found that the belfry was on the south end of the building exactly opposite its former position.

Down on Iowa avenue the tornado deposited a large telegraph pole in a place where telegraph poles are not usually seen. The owner of a house in the neighborhood had been repairing a chimney early in the afternoon and had left the scuttle-hole leading to the roof from the second story porch uncovered.

The tornado saw its opportunity and seized it.

It picked up a telegraph pole in front of the house, carried it over the roof and set it down squarely in the scuttle-hole. The pole was there for several days after the storm, its base resting on the porch and the cross-arms protruding through the opening in the roof of the porch.

HOW THE BIRDS FARED.

A dead robin was picked up in Lafayette Park. On one side the bird was intact. On the other every feather was gone. It was "naked as a picked bird," to use a familiar expression.

Instances have been recorded of the dove, the most timid of birds, seeking refuge with man when in great peril. Such an incident inspired the writing of the hymn,

"Jesus, lover of my soul,
. Let me to thy bosom fly."

A dove pursued by a hawk flew into an open window and into the bosom of John Wesley. It was that which inspired him to write one of the most widely known hymns of the century. During the storm a similar incident occurred. A number of people had taken refuge in the large corridor of the Union Trust Building. A snow white dove, unable to cope with the wind in its native element, equally unable to find security on any resting place exposed to the wind, blown hither and thither, in its fright entered the open doorway and flew to the shoulder of a gentleman standing there. The frightened bird seemed to know it was safe, and sat there quite and contented.

Henry J. Ruck, who lives at Thirteenth and Geyer, had a red bird and thereby hangs a tale. Before the storm the cage containing the bird hung on a nail on the back porch. Most of the porch was carried away. Mr. Ruck, in the general confusion, forgot all about the bird. An hour after the storm he looked out of the front window. The cage was lodged in the telegraph wires in front of the house and the red bird was singing gaily as if to say, "never touched me."

SOME CURIOUS EFFECTS.

Until repairs were commenced, the curious effects of the wind on certain objects was a source of much comment. At 1914 South Broadway, was J. B. Steffen's furniture store, which occupied a three-story brick building on the east side of the street. When the wind struck there it shaved off the entire front wall of the third story, but discriminated at the second. On each side close to the side walls, was a full length window with large panes of glass, one to the sash. All the rest of the front of that floor is gone, but these two windows are still in their places, the frames intact, and the glass not even cracked. The window to the south was evidently opened slightly at the bottom, as the drawn curtain behind it has been stained as if by rain, but otherwise no damage whatever was done to it.

The very tall steeple and its prototype of a minaret, which tower above the Catholic Church of SS. Peter and Paul at Eighth and Allen avenue, must have been lined with steel to have withstood the shocks as well as they did. The intermediate roof between them was badly wrecked, most of the debris falling into the body of the church, and the very steep roof of the east tower was also torn up very considerably. That the main roof should have been so damaged while the spires escaped is the more remarkable, as the church lies east and west, the direction of the storm's path, and hence the roof was to a very large extent protected by the two spires.

One of the out-of-place objects seen hanging to one of the spikes of Lafayette Park fence was a new spring bonnet. It was not claimed.

A coat and vest from some unknown source came sailing through the air and landed on the corner of St. Ange and Park. A negro soon after the storm appropriated the garments and was seen to abstract a "roll" from the vest pocket. No inquiry was made after for either clothes or money.

At one house a shirt was found under the back cellar door. As the cellar door was fastened down the only way the shirt could have got there was by blowing through the front cellar window. It was made for a man as big in girth as Grover Cleveland, and as the newspaper man is a light weight, the find did not profit him. He also acquired three trees, an old rug and an empty chicken coop, which latter was identified as the property of a butcher near the City Hospital, a block away.

On St. Ange avenue, from Park to Carroll, not a house,

UNION DAIRY COMPANY.

save one, escaped damage, many being entirely wrecked. All of them were occupied but one, and, strange to say, the empty house was the one which escaped. It was only slightly damaged by falling bricks from a chimney next door.

THE GREAT CYCLONE. 367

A POOR REFUGE.

Very embarrassing was the plight of three ladies who were on an open car of the Fourth street cable when the storm came up. For want of a better refuge they remained where they were. But when the unprincipled wind began to take unwarranted liberties they braved it and sought cover. The wind got into their big sleeves and blew out their skirts like balloons, until seams began to rip ominously and presently sections of garments were carried bodily away. It got to a point when to remain longer meant that they would be unclothed entirely, and they escaped while there was yet time. But in those awful few minutes even Peeping Toms had something else to indulge their prying propensities, and the ladies had no cause to blush.

CRAWLED UNDER THE BRIDGE APPROACH.

The Purina mills, Robinson & Danford, proprietors, situated just south of the Twelfth street bridge, was the scene of many narrow escapes. The full working force were in the mill at the time, including two young women who were in the third story. When the upper portion of the mill went down the employes ran out and crawled under the approach of the bridge, where it kept them busy dodging boards from the elevator on the east and brick from their own building, which were carried under the bridge with great force. The two young women in the third story did not get out until the roof had fallen, but fortunately they made their way safely to the bridge approach, where they also found refuge from the warring elements. A pony which was tied in the stable back of the mill, after the storm was found cutting all kinds of didos, and could not be pacified until Mr. Robinson found in its stall an elevator belt chain about 9 feet long which had been blown in from the elevator east of the bridge, a distance of about 400 yards. When the chain was removed the pony became quiet.

ALONE WITH HER BABY.

Mrs. L. Schnute, who lived at 838 Gratiot street, was alone with her babe. Their two-story residence is a complete wreck down to the first floor, furniture and all. Mrs. Schnute took her babe when she saw the storm coming and stood in the hallway close to the door. Where she stood is a space about 6 feet long and 5 feet wide not molested by the wreckage. Anywhere else in the building they would have been killed. Mrs. Schnute says she owes her rescue to James Stevens and Ed. Scanlon, the latter a watchman for the S. N. Long Syrup Company. Mr. Scanlon was blown from the third story of the S. N. Long Syrup Company building and the soles of his shoes torn from his feet; otherwise he was not injured, and is credited with many noble deeds of rescue.

MRS. WORHEIDE'S EXPERIENCE.

Ever since the cyclone Mrs. Fred. Worheide has been wondering how she and her two children escaped death. Her home is just across the street from St. Henry's Church, corner of Hickory street and California avenue, which was so badly wrecked. She and her little son and

THE GREAT CYCLONE. 369

daughter, 5 and 7 years of age respectively, were on the second floor when the cyclone swept down upon the residence. The entire second story was lifted up and carried away and a moment later her little girl was drawn away by the force of the wind and dropped upon the debris on the street. Mrs. Worheide and her son escaped with slight injuries, but the little girl, it is feared, may not recover.

Mrs. Worheide says the wind seemed to hold her fast, while her little daughter, who was standing only a few feet from her, was blown upon the street. Judging from this, it is thought there must have been several twisting counter winds sweeping through the building.

AN ADVENTUROUS AND COSTLY RIDE.

Ferdinand F. Herold, of the Herold Livery Co., No. 1717 Park avenue, sent Mr. Jerome Hill a bill for $155 for a carriage on the night of the cyclone.

"I did not charge half what I should," said Mr. Herold. "I am a wreck myself and two horses are ruined. I went off and left my family in the woodshed and one of my men is half crazy. I would not go through it again for $1,000.

"Mr. Hill came to me five times and begged me to hitch up and take his family down to the hotel. The office was full of injured people, waiting to be taken home. Mr. Hill tried to get Mr. Keyes to take him down and Mr. Keyes told him he would not hitch up for any amount of money.

"Finally Mr. Hill asked me if I would hitch up if he

got a lantern. He offered a man across the street $10 for a lantern before he got one. I hitched up myself and

EAST OF LAFAYETTE PARK.

had to drive over the sidewalks most of the way, and I had to walk ahead and cut wires or lift them up every few steps. I cut thousands of wires with pincers, and some I had to cut with a chisel. All this time I was wringing wet. On the way down to the hotel we were stopped every block by the police, who told us we could not go on. I told them not to bother me, that we could go on, and I did go through where it was thought no one could pass. I was wet all this time and shivering with the cold, and it was just as bad coming back. I left one man, his name is Gustav, on the other end of the Eighteenth street bridge. He is out of his head yet, and I don't know if he will ever get well again.

"Mr. Hill asked me to get him to the hotel, no matter what it cost, and I did it, and the bill I sent him is a very moderate one. I would not go through it again for any amount of money."

BROKE UP A WEDDING.

On Wednesday, May 27, a marriage license was issued from the office of the County Clerk at Belleville to Robert R. Haig of Caseyville and Miss Eliza Collins of Birkner.

The wedding was to have taken place that evening and the wedding party was assembled at the home of the bride, when the mighty wind swept down the hillside and scattered death and desolation on every hand.

The Collins home, like all the rest, went down, and many of the wedding guests were injured, including the sister of the bride, but the bridal couple escaped without a scratch.

They were taken to Belleville on the special train and

sheltered by friends, and the thought of marriage was temporarily put out of their minds.

When the expectant groom again bethought him of the license, the precious document could not be found. The wind which broke up the wedding party had made its work complete by carrying away the legal warrant for it. The couple went to Belleville later and a duplicate was issued to them without extra cost.

EIGHTY-FIVE MILES.

There were just 85 miles of streets obstructed by debris as a result of the tornado. All of these streets were cleaned up in 10 days. In St. Louis there are 480 miles of improved streets, so it can be seen to what extent the tornado interrupted traffic. The storm district, on its outside lines, extended from the river on the east to Tower Grove avenue on the west, and from Olive street on the north to Lynch street on the south. The extreme limits were three miles east and west and two miles north and south.

An adequate idea of the damage can be formed when it is stated that if all the houses damaged and blown down were on both sides of one street, that street would just be 85 miles long. This figure is given by the Street Commissioner, who has been through the entire district, and from the reports of his general superintendent, who has cut a roadway through the 85 miles of streets.

In addition to this, there were many more streets obstructed by wires and the like, which are not counted in the total, as the obstruction was but slight. These streets were cleaned up roughly the first day. It took just three days to cut out a passage through the debris so that traffic could be resumed. As 1,000 men were employed at this work, the magnitude of the task can be realized. There are 90 alleys obstructed. These alleys represent about 20 miles. The street gangs are now engaged in cutting a passage way through these alleys. It is a very difficult task, as in some of them there are three and four feet of brick and other debris.

The work of repairing and rebuilding in the wrecked district was pushed as rapidly as possible. The force of the Building Inspector's office worked with all its vim to see that unsafe buildings are properly and promptly repaired. No permit to reconstruct or repair demanded. The inspectors were simply seeing that the work is done substantially and safely. No one was asked to show a permit. Official red tape was tabooed.

A RIDE ON THE WIND.

On Thursday, following the great storm in St. Louis, there was found on a farm in the northwestern part of Fayette county, Ill., a piece of tar roofing paper fully a yard square. Sixteen miles west of Vandalia, near Smithsboro, a photograph of two young girls, presumably sisters, bearing the name of a St. Louis photographer, was found in William Defree's door yard the next morning after the storm. As St. Louis lies in a southwesterly direction, and the course of the storm being from the southwest to northwest, it is very probable that these articles were carried that distance by the wind.

A KINDLY DEED.

A pathetic story and a generous action came to notice. Mr. Sanford Snyder, 65 years old, who at one time was quite well-to-do, but of late years has suffered reverses of fortune, was coming home on a Sixth street car the evening of the storm, from Carondelet, where he had been

ADVANCE ELEVATOR "B," EAST ST. LOUIS.

seeking employment. The car was at Eighth and Hickory when the storm struck, and everyone on the car was more or less injured, Mr. Snyder being knocked insensible. Someone carried him to the sidewalk, where he lay in the driving rain for some time. Finally recovering consciousness he returned to his home in a pitiable condition, only to find it destroyed, and his wife huddled in a corner of the only room in the house which had not been wiped out of existence. Since then they have had a severe struggle to exist. Their condition was learned by a conductor on the Laclede avenue line, who immediately appealed to his fellow employes for help in relieving the aged couple. A sum of money was quickly raised and taken to Mr. Snyder, where it was joyfully received. Mr. Snyder and wife were still living in the ruins of their former home at 1227 Merchant street.

ANOTHER LUCKY ESCAPE.

The mangled bodies of a large number of the people taken from the ruins tell plainer than words the horrible manner in which death was met, but at the same time a number of the escapes made by citizens are as miraculous as the deaths terrible. Each survivor of the tornado has a story to tell, and chief among them is Tim O'Connell, a track walker on the Belt Line, who is spending his spare time in thanking fortune that he and the members of his family were born beneath lucky stars. O'Connell was walking along the Belt Line track when the tornado caught him. The tornado lifted him into the river. As O'Connell fell he caught a rope which was attached to a pier, and managed to hold on in safety while the storm blew overhead.

At the home on North Missouri avenue were Mrs. O'Connell and her four children. When the storm struck the place it carried everything away, and left the family unscathed.

THE CHINESE AND THE STORM.

Little, if any, mention was made of the Chinese who were injured or the loss of property they sustained. No less than seven were injured, one of them seriously, while the loss to their laundries alone will, it is said, amount to more than $3,000. The more fortunate members of the race have not been idle, however. They are not only taking care of their sufferers, but have raised $100 for the benefit of other unfortunates.

Wong Chin Foo, the well-known interpreter, stated that a meeting of the Chinese merchants was held and measures were at once taken to alleviate their countrymen's sufferings. The injured were given medical attention by Chinese physicians, while those who lost their property have received substantial assistance financially. They had little difficulty in raising the $100 for all other victims of the storm.

HENRY WEDERMEYER'S EXPERIENCE.

Henry Wedermeyer, of 911 Warren street, a switchman in the employ of the Wabash Railroad Company, was at work in the yard of that company when the storm began. "About half-past 5 o'clock," said Mr. Wedermeyer, "I went over to the buildings of the Belcher Sugar Refinery and climbed in a tank in one of the old sheds to take refuge from the storm. As I stood there pieces of flying boards and broken glass fell all around me. While looking south I saw the cupola of the St. Louis elevator and the biggest part of the roof blown off. The old buildings of the Belcher Sugar Refinery, between

Dixon and Ashley and Lewis and Main streets, were nearly entirely unroofed. The conveyors, about eight in number, on the river house of the St. Louis elevator were partly destroyed, and the top of the building where they weigh and shift the grain from one bin to another, was blown off. The smokestack on the St. Louis elevator was demolished. The St. Joe House, on Lewis street, between Ashley and Biddle, was totally unroofed and the gable end of the building was blown in. A box car on the Merchants' Elevator tracks, foot of Mullanphy street, was blown off the track and landed on its side. The Merchants' elevator building was also unroofed and badly damaged. The steamboats Polar Wave, Benton, Charlotte Boeckler and Jack Frost, lying between Smith and O'Fallon streets, were damaged, and the pilot houses and smokestacks being blown off and the sides demolished."

UNEASINESS OF SHANTY-BOAT OWNERS.

Where are the fishermen and others who occupied the boat houses in the beach just below the elevator? was a standing question on the Island after the cyclone. It is known that there were twenty or more of these boats on the bank, many of which came in when the river was at its height. Few were known, as they had come from a distance, and, finding a good harbor south of the elevator, anchored their crafts. On the morning after the storm all of these late arrivals had disappeared, and the denizens of that locality now believe that the boats were swept into the water and the occupants were drowned. These boats were not as high up under the shelter of the elevator as the others, which may explain their loss; while the old-timers weathered the storm without a single fatality, and with little loss. No inquiries have been received at headquarters for lost fishermen or persons who

WRECK OF ELEVATOR ON THE LEVEE.

follow the river in boat houses for a living, but the residents of Sandy Hook firmly believe that when these people fail to write from the points along the shore designated by their friends, in the same manner as Gipsey letters are forwarded, a general howl will go up from different ports and a full investigation will be made. Few people can form any estimate of the vast number of persons who are born and reared in these two-room shanty boats. Every great city has its Little Oklahoma, and many smaller ones its Sandy Hooks.

All do not fish for a living, Many are first-class mechanics, and go from one town to another in dull seasons. They live cheaply on the water and pay no rents for land. A few are known who make fair livings selling fancy work; while others are gamblers and saloon-keepers. The estimate of the loss of twenty boats, with their inmates, is not considered too large by some persons; but there is no authentic record of the swamping of any of them. All may have silently left the mooring and crept down the stream to a better hiding place, even after the storm, unnoticed by any one.

STRANGE FREAKS OF THE WIND.

The wind played strange freaks.

A horse was lying dead on the track beside the locomotive of the wrecked Chicago and Alton train. "It seems to me," said a railroad man, "as if he had killed himself trying to help the locomotive pull the train out."

In the worst part of the wreck of the upper Louisville and Nashville offices a dog lived to bark his delight at being released from the chain which kept the wind from blowing him away.

One of the dead horses on the Island roadway was apparently stabbed to death with splinters.

A stable which stands in the roadway was pinned through the corners by wooden beams, which were forced into both sides by the wind. The ends of the beams, extending up and out from the stable, contained a pile of lumber carefully arranged, as if placed there by hand.

A TRUNK MYSTERY.

The police were worried by a trunk, which was fished out of the river at the foot of Elwood street, on the night of the storm by John Gamache, residing at 5901 South Broadway. A quantity of woman's clothing, some dresses, evidently belonging to a little girl, a child's picture and a large number of letters make up the contents of the trunk. The letters are addressed to Miss Emma Boyle and nearly all of them are signed J. W. Rice. They are written on letterheads of the Moser Hotel, St. Louis, the Ringo Hotel of Mexico, Mo., the Grant House of Rolla, Mo., and some are on plain paper. The address, 3531 Olive street, appears in one letter. In others Rice requests the recipient to address him at Post Office Station F.

One of the letters is signed by Charles Boyle, and from its contents it is evident that the "Miss" Emma Boyle in question is his wife. The letter is remarkable for the extraordinary spelling of several words. "Answer," for instance, is spelt in the delightfully original manner, "ancer." The writer refers to himself as having been a "devorsed" man, and declares that he loves Emma as "mutch" to-day as he ever did. This letter opens with the single word "Mildred." It was evidently written by Boyle to a woman whose first name he uses in opening. Apparently she gave the letter to the owner of the trunk, who is thought to be the Emma Boyle before mentioned. In this communication, Boyle discusses at length his wife's actions and certain threats which she had made about leaving him. He declares himself at a loss to account for her coldness and says he has always treated her right.

The letters from Rice are addressed to the general delivery, and not to the house address of the recipient.

It is thought possible that the owner of the trunk perished in the storm. Theories that it was blown from a train on the Eads bridge or a wagon on the roadway of the bridge are also advanced.

HAVLIN'S THEATER DAMAGED.

About two-thirds of the west wall of Havlin's Theater building fell into the alley with a crash during the first storm. The section of the wall which fell left the entire stage exposed and the scenery and canvass decorations were loosened and torn down. The rent in the wall extends from the roof to the ground. The walls which have remained standing are bulged and cracked in places. A family residing on the third floor of the building, thinking the entire structure unsafe, sought other quarters.

ENGINE HOUSE COMPLETELY DEMOLISHED.

Engine House No. 7 was completely demolished by the storm. The hose reel and the horses belonging to it were buried under the debris in the cellar. The engine had responded to an alarm of fire at Ninth and Walnut streets, and escaped destruction. At the time the crash came five members of the company were in the engine

ALL THAT WAS LEFT OF A BEAUTIFUL HOME.

house. The sound of the tower in the rear of the house falling gave them timely warning, and they ran out and escaped injury.

Joseph Utley and "Tip" Lanham, of Engine Company

No. 7, lost their homes and all their household belongings by the storm, though their families escaped injury. The men have been kept hard at work since the calamity rescuing people from ruins and recovering dead bodies.

KILLED ALMOST OPPOSITE THE MORGUE.

Josephine Martini, 19 years old, was almost instantly killed in her father's shop, at 402 South Twelfth street, by falling debris. She was attending the business. It is a lunch stand, the dimensions of the room being about 4x12 feet. The flashes of lightning and peals of thunder frightened her and she attempted to run to the yard to the family's living apartments. As she reached the rear door the little dilapidated structure caved in and she was buried beneath its contents. Joe Diggs, porter at the morgue, dug out the lifeless body of the girl and with the assistance of the Superintendent of the morgue placed the body in an ambulance and had it conveyed to the Dispensary. Here the body was pronounced dead and removed to the Morgue.

The girl was killed almost opposite the Morgue. The mother of the dead girl, hysterical with grief, saw the body of her daughter placed on a slab and then she swooned.

PANIC AMONG LAUNDRY GIRLS.

The scene on Grand avenue, from the Fair Grounds to Carondelet, after the storm, resembled the thoroughfare of a besieged city after a heavy bombardment by the enemies' guns. The street was strewn from one end to the other with trees, whole trees and broken branches, broken telegraph, telephone and electric light poles, that made traffic of every kind both tedious and dangerous in the darkness and fallen wires, and it was with great difficulty that pedestrians wormed their way along the sidewalks, which were strewn with debris of every description. The middle of the thoroughfare was blockaded with electric cars of the various lines that traverse it from end to end, and a number of serious accidents occurred on account of horses becoming entangled in the wires that strewed the ground everywhere.

For half an hour after the storm struck the West End pandemonium reigned supreme at the Excelsior Laundry, corner of Grand and Bell avenues. Over 125 women are employed in this establishment, and the first gust of the storm crashed in the windows of the building, filling the air with broken glass and wrecking everything that was not nailed down, throwing the entire establishment into confusion.

Many of the women were slightly injured by the bits of flying debris. All made a rush for the exits from the building and rent the air with their screams, and several who fainted were nearly trampled to death by their companions in their frantic efforts to escape. The damage to the building is inconsiderable, and with a few exceptions the women were much more frightened than hurt.

All the streets in the West End presented a scene very much similar to that on Grand avenue, and in many places the streets were almost completely blocked with fallen trees, fences, poles and other debris. Chimney tops, window blinds, sign boards and broken glass were everywhere, and the intense darkness and pouring rain, lighted up by frequent and vivid flashes of lightning that blinded and terrified the soaking and dripping multitudes of working people who were compelled to grope their way home in the darkness and rain, made a scene that defies description.

A PRIEST'S ESCAPE.

Rev. Father Head and his sister narrowly escaped with their lives. Both are painfully injured, especially the reverend father, whose injuries are of a most dangerous character. The tornado tore off the two towers of Annunciation church on Sixth street, near Chouteau avenue. These with the heavy bell of the church fell with a crash onto the three-story residence of the priest immediately north of the church. They smashed through the house from roof to cellar. Father Head and his sister, who were in the room on the second story, were thrown into the cellar, as was their housekeeper.

The injured ones were carried as soon as possible by tender hands to the residence of Dr. J. C. Lebrecht at Sixth and LaSalle streets, where they spent the night.

Father Head had a terrible scalp wound six inches long and another in the back of his head. Both of his eyes are injured. His left knee is violently sprained and his right thumb crushed. He sustained a shock to his system that may prove very serious.

Miss Head, while not so badly off, had both hips injured and is internally hurt.

The housekeeper escaped with slight injuries.

THE GREAT CYCLONE.

DIED IN THEIR ARMS.

One of the most affecting stories of the terrors of the tornado comes from members of the Fire Department who worked on the wreck of Anchor Hall, at the south-

IMPERIAL LAUNDRY.

west corner of Jefferson and Park avenues. Down under the debris lay the body of poor Joseph Meyer, a barber employed in the shop in the building. When the storm came up he attempted to take refuge in the basement, but was caught in the crash before he had reached a place of safety.

The entire two upper floors came down upon him, and he was pinioned in the most painful position imaginable. He could scarcely move hand or foot, and the pressure of the tons of wreckage above him was gradually crushing his life out. He fully realized his position, and knew that unless assistance reached him within a short time he would perish. He called to the men he heard above him when the storm abated, but could not make them hear.

It was known that Meyer was under the wreck, and as soon as possible a rescue party was organized to search for his body. For long, weary hours he waited and listened, and as soon as he heard them coming, he called out the exact position where he lay, and rendered all the assistance his enfeebled condition would permit. As his rescuers gradually drew nearer he encouraged them and begged them not to give up the task. But it was slow work digging the mass of broken timber and brick and mortar out of the cellar, and day dawned before the rescuers came within speaking distance of the unfortunate.

During the long dreary night not a word of complaint came from the dying man, and not once did he chide them for not working faster. He occasionally showed the torture he was undergoing by groaning, but whenever they called to him to know if he could hold out till they reached him his answer was invariably that he was all right and that he would keep his senses about him long enough to direct the movements of those who were working to save his life. He said he wanted to see the members of his family and would not allow himself to think of giving up when help was so near.

At last about nine o'clock in the morning, the last stick of timber was removed from the dying man's legs and he was carried to the sidewalk above. He smiled faintly on his rescuers as they bore him out of his prison, but in spite of his efforts to appear strong and brave it was plain to be seen that he could not last long, and almost in the arms of the men who had come to his relief he breathed his last. Before he died he said: "Well, boys, I did my best to hold out, but I guess I'm going now."

HERO OF BALAKLAVA DEAD.

Among the victims of the tornado was Thomas Griffin, ex-policeman, ex-soldier and at one time a member of the famous Six Hundred which made the equally famous charge at Balaklava. He and three daughters were caught like rats in a trap at their home on Seventh street, and were crushed beneath the walls. Griffin was one of the best known men in the city, and served on the police force for nearly eight years. During this time he was stationed at the approach to the Eads Bridge and was known by sight to thousands of strangers who visited the town. Griffin served with the Light Brigade during the battles of Balaklava, Inkerman, Sebastopol and Lucknow. At the time of his death he was wearing medals received for bravery in each of these battles.

FOUND SAFETY IN A BASEMENT.

The Wetzel family enjoyed the grandeur of the storm until the windows of their home began to give way. They saw the first chimneys of the Union Club fall, and the running comment at once ceased. Each member of the household rushed to the basement, and no sooner had they congregated than the walls of their home gave way. The loss of furniture was complete and the house a wreck.

Dr. William A. McCandless, who lived in a handsome white stone residence of three stories adjoining the Wetzels, looked at the wreck of his home, and remarked: "I will give all that is left to anybody who will remove the wreckage."

The McCandless household was saved by taking to the basement, for all that remains of their home are the walls.

HOW THE HORSES BEHAVED.

At 2710 Lafayette is the ruin of a mammoth livery stable, owned by Walter S. Gregory. The entire second floor is gone and 15 carriages are destroyed. When Mr. Gregory told of his loss yesterday his affection for the horses which had escaped unhurt was clearly apparent.

"They knew as well as human beings could have known," said he, "that death stared them in the face. They were not unruly. Their piteous neigh revealed their fear, and they were as docile as cowed school children would have been."

In this ruined barn lay the remains of Mrs. F. D. Woodruff, wife of a barber at 2743 Ann avenue. When the storm struck her home Mrs. Woodruff was on the second floor of the building. She hastened to the rear of the house, and clutching a clothes line, fastened it around her body. But before see could throw herself from the window the walls gave way and she was hurled to her death.

Diagonally across the street from the Gregory stables

FOURTEENTH AND PAPIN STREETS.

are the stables of the Compton Avenue Livery Company. Here but little damage was done to property, but in an attempt to hold the doors against the storm, Joseph Warren, the foreman, was struck by a piece of flying glass and an artery was cut in his neck. He was not expected to live.

A freak of the storm was to lift the frame of the elevator shaft from the Compton stables and carry it a distance of 150 feet, when it was deposited over a telegraph pole. The direction taken by this shaft was southeast, something which cannot be explained according to any statements yet made as to the direction of the wind.

MRS. EYERMANN'S MIRACULOUS ESCAPE.

By far the most miraculous of all the escapes anyone had to tell was the story related by Mrs. Dr. C. H. Eyermann of 1722 South Jefferson avenue. The Eyermann's lived in a two-story brick containing nine rooms. The father was away from home at the time and the mother with her nine-year-old-son was on the second floor. The windows in the second story blew in and Mrs. Eyermann went to the first floor and took her son with her. The walls in the second story began to fall and she went to the basement. She had left the gas burning in the main hall and cautioning her son to remain quiet hastened above to put out the gas, for fear the house might be set on fire.

No sooner had she turned out the gas than the whole house seemed to be giving way. She rushed down the stairway leading to the basement and caught her child just in time to force him between two pieces of masonry that formed a small archway underneath the hall. The brick house fell with a crash. Not a brick was left in position, and somehow, in some way, "God only knows," remarked Mrs. Eyermann, neither the mother nor the child was injured.

Their next endeavor was to extricate themselves. Above and all around them the falling bricks had gathered, but they followed the archway to the main basement and finally dug away enough bricks from around a window to gain their freedom.

Dr. Eyermann was only a couple of blocks from home in his buggy, but was seriously injured.

Mrs. Eyermann sat on an improvised seat made by resting a board on the bricks from her home the next day and complacently ate supper.

GAVE HIS LIFE TO SAVE A HORSE.

There were many marvelous escapes along that part of Missouri avenue which faces the park. The home of Alois Solerer is a three-story brick, with imitation stone front, at 1401 Missouri avenue (corner of Park). The roof is blown away, and the family was gathered in the basement, while the brick and mortar fell thick about them. In this family group was also William Taylor, the negro coachman, who had been with the family for years. The negro was passionately devoted to "Bess," the family horse, and while the storm was at its worst heard the animal "neigh."

It was the same piteous neigh which has already been described by the proprietors of several stables and Taylor started for the stable. The family pleaded with him in vain. Taylor reached the stable just in time to free "Bess" just before the walls gave way. Although the horse escaped the negro was buried in the ruins, and was taken out dead.

There were some stories of miraculous escapes to tell at each of the houses facing Lafayette park from Missouri avenue, but the general tenor was the same. The people escaped by fleeing to the basements of their houses.

Looking east from Park avenue the wreckage seemed complete. Save the one residence block which bounded the park on the east the others were commercial and industrial streets. The small shop keepers along Park avenue, east of the park, were with their families in a sad mixture of children, mattresses, broken bureaus and bedsteads and disfigured homes. The most colossal wreck of the neighborhood was the big two-story, brick car shed of the People's railway line. It covered one full acre of ground, and the whole structure fell, mashing the cars and building into a gigantic heap of brick and mortar. It was almost impossible for one to pick his way along Park avenue, and the mass of humanity which struggled to cover the territory thereabout included every variety of citizenship which a big city has to offer.

Small boys rang the gongs and bells on the despoiled street cars, linemen vainly attempted to rescue such electric wires as could be preserved.

"SAVED THROUGH GOD'S MERCY."

No one will ever satisfactorily explain how the families of Edward and William Rea, living across the street at 2345 Park avenue, escaped. Their former home was a handsome brick, three stories high. Four women were in the parlor when the west wall gave way and filled the room with brick and mortar. To look at the room and imagine that any one of its occupants could have escaped is impossible. When asked how it occurred, Mrs. Ed

BROWN TOBACCO COMPANY'S BUILDING.
(FOUR PERSONS WERE KILLED IN THE WRECK ON THE RIGHT.)

Rea said: "We were saved through the mercy of God."

Mrs. S. Collins at 2343 Park avenue is equally at a loss to account for the escape of herself and four daughters. Before any one of the family attempted to take refuge in the cellar the rear walls to the house gave way, and the furniture was scattered in all directions. Through the flying debris they made their way to the cellar, and no one was injured. Their home and furniture are totally destroyed.

STRANGE BUT TRUE.

The tornado developed hundreds of incidents so unique that the best of them are entitled to a chapter to themselves. Since the account of the first of these strange storms was written, back in the days of the Marshfield and Grinnell disasters, stories of miraculous, unaccountable escapes, peculiar deaths and fantastic doings of the elements have formed a large part of the history of every storm of this character.

For example, nothing is more difficult of explanation than a condition which exists at the wrecked home of Dr. Starkloff on Compton avenue. The outer walls of the splendid red mansion are torn away, the roof is gone and there are other evidences of the ravages of the storm without the building. Yet the light pictures on the walls are hanging in place and the lamps on tables and stands are not disturbed as to position, neither are they in any way damaged. On one of them the delicate lace shade is not even disarranged.

In South St. Louis there is a house whose entire north wall is torn out save a support under one of the windows and the window itself. The frame is not damaged and not one of the panes of glass is broken.

On Russell avenue, not far from Compton, one of the

heavy marble steps that were in front of the main door of the residence was picked up and the end driven into the ground to a depth of two feet. The step is not chipped or in any way injured.

There are two iron posts in front of the Merchants' Exchange building. One of them was wrenched off by the storm. Sixty feet away is a wooden post of the same size and height of the iron pillar, and it was in no way damaged.

Chunks of ice, presumably from the Wainwright Brewery, were found in the water in Twelfth street, just below the Shickle-Harrison Iron Works.

HORSE UNHARNESSED.

On Grand avenue, not far from Shenandoah, a horse was torn from its harness and thrown, upside down, in an excavation that is being made for a sewer. . The other horse in the team was tossed across the street.

A bed with its mattress in place was thrown from a house on Missouri avenue to the center of Lafayette Park. The pillows fell not far from it, but they were not the pillows belonging to this particular bed.

A child's chair was taken from one of the lamps at the gate leading into Lafayette Park on the south side. Part of an arc lamp was fastened to it.

A middle-aged German was walking along the railroad tracks east of Twelfth street just after the storm, bemoaning the loss of his little daughter. Another child, who was with him, found a bonnet that was recognized as belonging to the missing girl. The father thought that it was conclusive evidence that his other youngster was lost when the little one ran up and shouted that she wanted to get in the house, because it was getting too wet for her.

Bales of hay ought to be good for defense in a cyclone. In many livery stables on the South Side everything but the great stacks of hay in the lofts was blown away. In one case carriages were taken half a dozen blocks and set down with little injury. The hay was not disturbed.

PET CAT FOUND.

A pet cat, owned by a family on top of Compton Hill, was found in front of the Lafayette Avenue Methodist Church at 9 o'clock Friday morning. It was not injured.

In a house on Arkansas avenue a glass stopper in a whiskey decanter was broken off at the top of the bottle, but the vessel itself was not broken.

A young man named Murphy of South Twenty-third street, says that he had two rings on his left hand. He lost them, with a good deal of the skin of two fingers, while trying to keep his hold on a telegraph pole. The pole was overturned and Murphy narrowly escaped being crushed under it.

Letters addressed to many prominent people were found in the trees of Lafayette Park. One was the property of Mrs. Charles Nagel.

The upper part of the stone house occupied by A. K. Hammond in Waverly place was torn away. A brick house in the same square was hardly damaged.

FRIGHTENED CHICKENS.

A chicken coop with two live and very much frightened chickens was found floating in a new-formed pool of water in Lafayette Park.

Two young women who were on a Fourth street cable

TRI-ANGLE WAREHOUSE.

car at the time of the storm were entirely denuded. They left the car stark naked and were cared for in a house near by.

Black mud was found on the walls of many houses after the storm. In one residence on Arkansas avenue the ceiling is decorated with black patches. There is no such mud in the neighborhood. Where did this come from?

There was a new spring bonnet on one of the spikes of the Lafayette Park iron fence. The bird on it wasn't stripped of its feathers.

The irony of fate was never more forcibly illustrated than in case of a dwelling-house located on South Broadway, the entire front of which was swept away, leaving the interior with furniture exposed to the elements. The furniture and bedding were piled in a confused mass upon each floor, while upon the rear wall of the second story bedroom was the legend, "Good Luck."

EXCHANGE MEMBERS TIMID.

There were great fears that the storm which unroofed the Merchants' Exchange had made that building unsafe, and at the opening the morning after the cyclone many of the more timid were fearful of entering. President Spencer at once sent for James Stewart & Co., the architects, who made a thorough examination and reported that there was no danger, everything being perfectly safe. A notice to this effect was posted upon the bulletin board, which had the effect of reassuring the members.

A large number of messages from various cities throughout the country were received by members of the Merchants' Exchange offering sympathy and assistance. Vice-President R. S. Lyons of the Chicago ·Board of Trade wired President Spencer as follows: " The Board of Trade of the city of Chicago is shocked by the horrifying calamity which has befallen your city, and extends its sincere and warmest sympathy." A cable received by the E. B. White Grain Company, from Alexander & Co., of London, England, said: "Our sympathy extended."

A FLYING STOOL AND A DOLL HOUSE.

A tall office stool fell from the top of the tower of the McLean building. It had no mark indicating whence it had come.

Ben Selkirk's house on the north side of Park avenue collapsed. With strange vagary, the wind blew pieces of brick, limbs of trees and pieces of slate through the parlor windows in front, knocking the chandelier into smithereens, and treating handsome vases and mantle ornaments in the same rough way. With singular freakishness, it missed a charming little doll house, which Mr. Selkirk had made with his own hands for his children. This doll house stood, and stands now, between the two front windows on a table. It is as good as new, and not even a drop of water got on it to mar its whiteness.

THREE LITTLE ONES DIED.

A touching incident occurred at the Bethesda Home during the prevalence of the tornado. The home is at Twelfth and Hickory in the line of the storm, but escaped uninjured. It was rocked and badly shaken, but none of the inmates were injured. Sad to say, however, three children died during the storm. Some think the deaths were caused by fright, but others say the children were dying before the storm broke. The death scenes were impressive. Outside the thunder roared, the lightning flashed, and crash after crash of thunder shook the home. The little ones looked into the faces of the nurses, appeared at first startled and then smiling peacefully they passed to the other world at the same moment other souls were being hurled into eternity by the fierceness of the storm and the unmerciful strength of the wind.

BLOWN INTO THE RIVER.

Jim Murray, employed on the Anchor Line wharfboat, was sitting on the wharfboat when the tornado descended on the Levee. Foreseeing the danger, Murray made a run for the shelter of the elevated road. The wind gratified his desire to seek this shelter, but not before it had some fun with him. Murray was lifted off his feet and blown over the "apron" of the boat into the river, landing in a dry dock moored close by, used by carpenters to repair the hulls of vessels. The next instant the dry dock, which is a hollow affair about 10 feet wide by 15 feet long, was blown westward out of the water, tearing off a portion of the railing of the "apron." It was driven with great violence against the iron supports of the elevated railway, dumping Murray out unceremoniously upon the ground. The dry dock was again taken up in a return current of wind and carried out towards the river almost to the water's edge, where it was caught by a reverse current, whirled high into the air and dashed to pieces against the roadbed of the elevated, scattering debris all over the wharf. Murray was dazed and pretty sorely bruised, but not injured otherwise, and he held on with a death grip to the Terminal elevated support until the storm had spent itself.

PAPER HANGER'S EXPERIENCE.

During the height of the storm Wednesday a paper-

WRECKED AND BLISTERED.

hanger named Stewart was decorating the walls of Ed Morrissey's saloon, opposite the Four Courts. When the front walls of the restaurant blew in and scattered debris through the house Stewart fled to the cellar. When his two assistants decided to follow him a few minutes later, they were convulsed with laughter at observing him stretched prone upon his back in the damp cellar, the water reaching nearly above his arms, with two heavy stones across his breast, which he had placed there to prevent the wind from blowing him away. Despite the "jibes" of the other occupants of the house Stewart continued to occupy his uncomfortable position until assured that all danger was past.

SAVED HIS DRINK.

Henry Collins of East St. Louis, tells a queer story of his experience during the storm. He was standing in a saloon in East St. Louis with a glass of liquor in his hand. Suddenly the roof fell in, he was turned over twice or thrice and landed on his feet with the glass still in his hand and half of the liquor still in it. He quaffed the liquor with relish, as his collarbone had been broken in the crash and he needed the stimulant.

WEDGED IN A CAR ROOF.

Mrs. R. P. Tansey, wife of the president of the Merchants' Transfer Company, well known in St. Louis social circles, but now a resident of Springfield, Illinois, was a passenger on the Chicago and Alton train that became tangled up with the tornado on the eastern end of the Eads Bridge. When the coaches turned over Mrs. Tansey, in some way, became wedged in the roof of one of them and there was considerable difficulty in extricating her; as it was, one of the brakemen finally pulled her through a window and she then footed it over the crossties in that overwhelming downpour of rain to East St.

Louis and tried to find the Martell House, but only succeeded in reaching the spot where it had once stood. How she ever got to St. Louis she does not know, but late in the evening she made her way to the Planters' Hotel, more dead than alive, where her anxious husband found her the next morning.

DIED FROM FRIGHT.

In Clifton Heights an old lady living in a house in the rear of the residence of L. Haller was killed by the shock experienced when an uprooted tree was dropped through the roof. She was not struck or injured in any way by falling debris, but her nervous system had been so shattered by the intensity of the storm that the additional strain was too much for her.

IN A FIRE-PROOF VAULT.

There were lively times in the General Auditor's office of the Missouri Pacific system on South Seventh street. On the second floor of the building there is a very large fire-proof vault, and when the winds were raging the fiercest, fifteen of the clerks sought refuge in it. The building, however, began to shake and quiver, and the strong vault began to totter in sympathy with the movement, which brought forth a howl of anguish from the imprisoned men. It was then too late to run for their lives, and all they could do was to stay with the safe. Luckily, the building was spared and no one was hurt. This building is probably packed and jammed with clerical occupants more than any other building of the same size in this city, and when the roof was taken away by the storm, the panic grew into the proportions of a small sized riot of frantic, safety-seeking people.

TWO WOMEN'S ESCAPE.

When the steamer Dolphin was taken in hand by the wind, and before being consigned to a watery resting place, it was dashed against one of the piers of the Eads bridge, where the force of the storm held it for a moment. J. W. Schoeffer, of General Manager Ramsey's office of the Wabash was on the bridge at the time, and saw two women crawl quickly from the steamer onto a ledge of rocks in the pier, where they clung and huddled until a rope was thrown to them from above. The boat, a moment or two after the women had gotten clear of it, was again put in action by the elements, turned upside down and capsized in the middle of the river.

SIR CHARLES GIBSON'S TREES.

One of the prides of Sir Charles Gibson—a greater pride than even the title he possesses, but never displays—were the trees that surrounded his beautiful home just across from the southwest corner of Lafayette Park. Reaching Lawyer Gibson's house, after a walk up Lafayette avenue of 20 or more blocks, the beholder was almost accustomed to the sight of fallen houses, and so the fact that the grand old trees were torn up by the roots seemed more impressive than the fact that the beautiful home they shaded was a total wreck, and its usual happy occupants obliged to seek shelter elsewhere in the city.

There were stories told by the collection of rubbish, piled like the drift that waves of the sea wash ashore, that bordered Lafayette Park, particularly on the southern side. Bits of dainty millinery, lying in close proximity to broken hat boxes, told the tale of festive spring

ST. LOUIS WIRE MILLS.

and summertime headgear that must forevermore remain nothing but memories.

Tiny crib mattresses, caught in a tangle of shattered pieces of woodwork and fallen limbs of trees, hinted so plainly at the possibility of a wee bit of humanity being hurled without mercy through the air, that one hesitated to continue investigation for fear of uncovering suddenly a pale little face.

The bits of polished wood, with handsome metal handles attached, that still retained shape enough to be distinguished as a lady's writing desk, accounted for a littering of torn letters and envelopes, bearing the name and address, as closer observation disclosed, of Mrs. Charles Nagel. Of course, it goes without saying the desk was caught and blown from its place in the residence opposite the park, which is now among the ruined.

COLORED VICTIMS.

The colored people along Papin and Gratiot streets could be seen for blocks in the half open or entirely uncovered and unsided houses gathering together their belongings.

On the north side of Papin, Gratiot and Caroline streets nearly all the front walls are out, while on the other side of these streets all the way up from Jefferson avenue the people are saved the trouble of raising their windows. They were all leaning out of the empty frames, watching the slow and tedious clearing up process of their neighbors across the way.

WILLIE WINCKLER'S DEATH.

Poor Willie Winckler, nephew of Adolphus Busch, was in high glee all day Wednesday. Early in the morning he received a cablegram announcing the safe arrival in Kreuznach, Germany, of his wife, to whom he had been married a little over a year. In the evening he got on the Seventh street electric car to take the cablegram up to his mother-in-law, Mrs. Meyer, that she, too, might be apprised of the safe arrival of her daughter among her husband's people. He had been on the car but a little while when the storm broke loose. Mr. Winckler, the conductor told his relatives afterwards, was exceedingly nervous, and could hardly be restrained from jumping off the car. Finally he managed to swing off, the conductor still urging him to remain inside. Just then a terrific blow of the hurricane overturned the car, and Mr. Winckler was buried under it. When the debris was removed he was dead.

ACTS OF HEROISM.

Many heroic acts were performed in the saving of lives as a result of the storm.

When the City of Monroe was twisted away from the Anchor Line wharf, there were about 40 passengers on board and a full crew, as the boat was just making ready for the trip to New Orleans. When the moorings finally gave way the boat lurched over on its starboard side and nearly capsized. The movement threw nearly all the freight to the starboard side and this served to hold the boat in this perilous position. Captain Viegler made a reassuring speech to the passengers which slightly quieted the extreme excitement. He said they were all safe. When the boat struck the Illinois bank the captain was not to be found.

At 7:30 o'clock Wednesday night, at which time the last news from the boat was received in St. Louis, he had

THE GREAT CYCLONE.   409

not been found, and it is feared he was blown overboard. Fred W. Williams, a colored roustabout on the boat, saved four lives, all of children. Mrs. Mary Mannle, of 111 Collinsville avenue, East St. Louis, and her four children had been placed on the boat at 3 o'clock by their husband and father, Joseph Mannle. They were going to visit relatives at Cairo, Ill. Mrs. Mannle, with her children huddled close about her, was on the lower deck, awaiting, as she supposed, sure death. When the boat grounded, Williams, in answer to her appeals, grabbed one of the children and plunged overboard. A few strokes brought the little one safely to land, and Williams made three other trips, landing the other children on the river bank. He performed this act of heroism in the midst of the storm, when there was the likelihood that the boat would be capsized every moment. When the wind had spent itself, the other passengers were landed in safety. Mannle, the father and husband, was soon on the scene, and Williams was the object of general worship at the Mannle homestead that night.

SAVED HIS CAPTAIN'S SON.

On the Libbie Conger, which was ground to pieces on the Illinois shore, were Captain Seaman, his wife and his child, a little boy who is a general favorite on board the boat. When the worst appeared to be coming, a negro, whose name is unknown, took the child in his arms and plunged overboard bearing the captain's boy safely through the boiling waters to the Illinois shore. Captain Seaman rescued his wife. But for the act of bravery of one of his roustabouts his boy would probably be among the missing.

When John McDowell, a roustabout on the tug Reliance, which performed great work among the boats which lined the Illinois banks for miles down the river, returned

COMING HOME AFTER THE STORM.

to the city, he was burdened with messages to loved ones in this city from fathers and brothers telling of their safety. He brought happier news to many households than was ever brought to them before.

He reported that the Madill had been blown to pieces

and that the crew swam ashore. He said he thought every member of the crew had escaped with life. The ferries at the foot of Anna street and at East Carondelet had been blown to pieces and gone to the bottom with their crews and passengers. This could not be verified, but the negro stated positively that such was the case.

There was a rumor that the excursion steamer Grand Republic, belonging to the Columbia Excursion Company, had gone to the bottom with 500 excursionists. An officer of the company denied this. He said the boat left St. Louis at noon to go to Alton, where it was engaged for an excursion at 8 o'clock that night. The storm might have blown the boat away, but in such exigency only the crew would have been imperiled and these men could all have swam to safety.

CABBY WAS AN AUTOCRAT.

The hotels and lodging houses down town were crowded to their full capacity the night of the storm. The Olive street cable line was the only dependence for those who live in western part of the city. To the south there was no street car communication and operation of the suburban street car lines was suspended. There was no way for thousands of men and women who work in the city to get to their homes save by cabs, and every cab in town was busily engaged the greater part of the night, though driving in the dark streets was dangerous. Old express wagons, furniture vans and every other sort of vehicle was pressed into service during the early evening hours. Many anxious to get home and assure themselves of the safety of their families paid as high as $10 for a ride in a grocery wagon. Prices for cabs were correspondingly steep, and cabs were hard to get at that.

DYING IN A DRUG STORE.

Agonizing groans emanating from Leland Miller's drug store at about 9 o'clock in the evening attracted the attention of passers by, who soon learned of the terrible suffering of G. C. Poppitz, another tornado victim. He was walking aimlessly up Locust Street from Broadway at about 8 o'clock, and when he reached the corner on Sixth in front of the drug store, he dropped to the sidewalk unconscious. One of the clerks of the store assisted him inside, and restoratives were promptly administered.

The suffering man regained consciousness at times, but talked incoherently. He said he had walked from East St. Louis, but did not remember how he had crossed the river. Later in the evening he said he was struck in the side while riding in a car, but he could not recall what had hit him.

An examination of the injured man showed that he was fatally injured, his entire right side and ribs being crushed in and his lungs apparently compressed. He expectorated blood continually. The attending physician pronounced his case hopeless.

A carriage was dispatched to his home at 2501 South Broadway with a messenger to notify the parents of the unfortunate young man, but up to 10 o'clock nothing had been heard from them or the messenger.

Young Poppitz is a graduate of the College of Pharmacy, of the class of '96, and appeared to be about 21 years of age. His father is in the furniture business, being a member of the firm of Kluegel & Poppitz, whose store is at 2501 South Broadway.

A PANIC AVERTED.

Captain Pat Carmody's presence of mind prevented a panic at the Fair Grounds. When the crowd stampeded for protection from the tornado, Pat corralled the frightened people underneath the stand and forced them to remain there. He had a mob of about 300 in check, and it was lucky for them that he did so. Everybody was trying to get out and wanted to make a break for the south side of the stand. While Carmody was holding them back, the roof of the stand was blown off, and the debris fell just where the people wanted to go. Carmody's good judgment saved many lives.

Dan Honig, the well-known horseman, fainted away during the height of the storm. Dan is a robust-built man, but he is subject to heart failure, and his friends thought he was a "dead one."

When the stand fell R. J. Pearson's horse and buggy was caught in the debris and completely buried. The big posts saved the horse and he was all right when dug out. His escape was simply miraculous.

Drs. Bernays, Newman and Neville made a tour of the track after the tornado and found no one injured.

Hundreds gathered in the infield when the storm blew up and were thoroughly drenched. Frank James found a hill and located close to it. He ordered the people to lie down on the grass and everybody took his advice.

The roof of the stand was blown in all directions. Part of it landed on Natural Bridge road and several shanties were buried beneath the debris. The bookmakers left the betting ring in a rush and their cashiers flew with

SINGULAR FREAK OF THE WIND.

them. In the crush a number lost part of their bankrolls and the touts had a scramble for the coin.

Mrs. Charles Van Dusen, wife of the well-known jockey, was caught in the stampede and knocked down. She lost her pocketbook containing $110. Another woman from

San Francisco lost her purse, which, it is said, contained several hundred dollars worth of diamonds. The lady had a little child with her and the youngster behaved with startling bravery.

One of the supporting pillars of the stand fell through the first floor when the roof went off and a number had narrow escapes from being crushed to death. Johnie Goebel, one of the form book men, was near by at the time and had the narrowest escape of his life.

The awning over the club house veranda was blown off. It suffered the same fate last season.

Clayton Woods, who works for one of the books, was trying to get out of the betting ring during the height of the storm when a bettor asked him to pay off a winning ticket on one of the horses in the fifth race.

"Pay off!" said Clayton; "not for me. I guess we will all have to account this time to the main squeeze up above."

George Munson and Secretary Aull were hard at work trying to quiet people after the worst of it was over. Munson was drenched to the skin.

TRAIN'S NARROW ESCAPE.

While the storm was at its height, passenger train No. 7 of the Chicago and Alton road pulled out of the bridge from the Missouri side on its way east. Engineer William Swoncutt had only proceeded a short distance when he realized the awful danger which threatened the train. The wind struck the coaches at first, startlingly, causing them to careen omniously. At that time he was about half way across. Overhead telegraph poles were snapping and tumbling into the river, while several large stones were shaken loose from their foundations and came toppling down into the water. Fearing that every moment his train would either be blown into the water or else the bridge would be blown away beneath him. Swoncutt, with rare presence of mind, put on a full head of steam in an effort to make the Illinois shore.

The train had scarcely proceeded 200 feet, and when within about the same distance from the shore an entire upper span of the driveway of the bridge was blown away. Tons and tons of huge granite blocks tumbled to the tracks, where the train loaded with passengers had been but a moment before. At about the same instant the wind struck the train full on the side, upsetting all the cars like playthings. Luckily no one was killed in the wreck, but several were taken out severely injured.

The wrecked part of the bridge is just east of the big tower, near the Illinois shore, and extends east for about 300 feet. The entire upper portion, traversed by the cars and carriages, is carried away, while the tracks beneath are buried in debris, in some places eight feet high.

FOUND A DEAD BABY.

An unknown man who came into the Union Station at 7 o'clock to meet his wife, who was expected on an evening train, reported that he picked up a dead baby in the street at Twenty-second and Market streets. He gave it to a woman, who said she thought it was the child of a woman living near that corner.

TIM O'NEIL is a reporter for the *St. Louis Post-Dispatch*. He formerly worked for the old *St. Louis Globe-Democrat*, a number of other newspapers, and the Associated Press in the South. O'Neil grew up in St. Louis and graduated from the University of Missouri School of Journalism. He lives in south St. Louis with his wife and two children.